T0136876

PoliTO Springer Series

Series editors

Giovanni Ghione, Turin, Italy
Pietro Asinari, Deparment of Energy, Politecnico di Torino, Turin, Italy
Luca Ridolfi, Turin, Italy
Erasmo Carrera, Deparment of Mechanical and Aerospace Engineering, Politecnico di Torino, Turin, Italy
Claudio Canuto, Department of Mathematical Sciences, Politecnico di Torino, Turin, Italy
Felice Iazzi, Department of Applied Science and Technology, Politecnico di Torino, Turin, Italy
Andrea Acquaviva, Informatica e Automatica, Politecnico di Torino, Turin, Italy

Springer, in cooperation with Politecnico di Torino, publishes the PoliTO Springer Series. This co-branded series of publications includes works by authors and volume editors mainly affiliated with Politecnico di Torino and covers academic and professional topics in the following areas: Mathematics and Statistics, Chemistry and Physical Sciences, Computer Science, All fields of Engineering. Interdisciplinary contributions combining the above areas are also welcome. The series will consist of lecture notes, research monographs, and briefs. Lectures notes are meant to provide quick information on research advances and may be based e.g. on summer schools or intensive courses on topics of current research, while SpringerBriefs are intended as concise summaries of cutting-edge research and its practical applications. The PoliTO Springer Series will promote international authorship, and addresses a global readership of scholars, students, researchers, professionals and policymakers.

More information about this series at http://www.springer.com/series/13890

Letizia Lo Presti · Salvatore Sabina
Editors

GNSS for Rail Transportation

Challenges and Opportunities

Editors
Letizia Lo Presti
Department of Electronics
and Telecommunications (DET)
Politecnico di Torino
Turin
Italy

Salvatore Sabina
Ansaldo STS
Genoa
Italy

ISSN 2509-6796 ISSN 2509-7024 (electronic)
PoliTO Springer Series
ISBN 978-3-030-07723-5 ISBN 978-3-319-79084-8 (eBook)
https://doi.org/10.1007/978-3-319-79084-8

© Springer International Publishing AG, part of Springer Nature 2018
Softcover re-print of the Hardcover 1st edition 2018
This work is subject to copyright. All rights are reserved by the Publisher, whether the whole or part of the material is concerned, specifically the rights of translation, reprinting, reuse of illustrations, recitation, broadcasting, reproduction on microfilms or in any other physical way, and transmission or information storage and retrieval, electronic adaptation, computer software, or by similar or dissimilar methodology now known or hereafter developed.
The use of general descriptive names, registered names, trademarks, service marks, etc. in this publication does not imply, even in the absence of a specific statement, that such names are exempt from the relevant protective laws and regulations and therefore free for general use.
The publisher, the authors and the editors are safe to assume that the advice and information in this book are believed to be true and accurate at the date of publication. Neither the publisher nor the authors or the editors give a warranty, express or implied, with respect to the material contained herein or for any errors or omissions that may have been made. The publisher remains neutral with regard to jurisdictional claims in published maps and institutional affiliations.

Printed on acid-free paper

This Springer imprint is published by the registered company Springer International Publishing AG part of Springer Nature
The registered company address is: Gewerbestrasse 11, 6330 Cham, Switzerland

Preface

This book comes from the need to combine the topics of positioning, based on Global Navigation Satellite System (GNSS), with the application of this technology in the railway environment. Therefore, this book is aimed at readers who need to explore the technological implications of using GNSS in an application environment with safety of life (SoL) characteristics and, in particular, in the rail one that has properties very much different from the aviation and the maritime application domains. We assume that the readers are not expert on the specific topic of GNSS for SoL applications, but already know GNSS and the new policy on the use of GNSS in the railway environment, established by the European initiatives such as

- European GNSS Agency (GSA) (https://www.gsa.europa.eu/sites/default/files/futureflowchartd.pdf)
- European Space Agency (ESA) (www.space4rail.esa.int)
- Shift2Rail (https://shift2rail.org/)
- European Union Agency for Railway (ERA) (http://www.era.europa.eu/Document-Register/Documents/ERA-REP-150_ERTMS_Longer_Term_Perspective-Final_Report.pdf)

In this context, three key points can be identified: (1) Positioning information provided by GNSS must be accurate, safe, and reliable, rather than only extremely precise, (2) GNSS Positioning should be integrated into a railway signalling system so that the Train Position function still guarantees the required Safety Integrity Level (SIL) (i.e., SIL 4, the highest level) and the reliability, availability, and maintainability requirements and (3) GNSS Position should be provided taking into account the complex system of regulations governing railway signalling systems.

This book then develops into two distinct parts. In the first part, covered by Chaps. 1–5, the issue of GNSS integrity is addressed, which is the discipline dealing with the issue of reliability of position information provided by GNSS. The second part, covered by Chaps. 6 and 7, describes the main requirements, the methods, and the regulatory framework for railway positioning systems with focus on the introduction of the GNSS Positioning into the evolution program of the European Rail Traffic Management System (ERTMS). The introduction of satellite technology in this ERTMS environment is not state of the art yet, because the definition of a **standard**, **interoperable**, and **backward compatible** solution is still in progress. Therefore, this book cannot describe well-established and consolidated solutions, since they do not exist yet, but it provides a framework and tools to design and implement innovative solutions, in line with the (R&D) programs of the GNSS application in the rail.

Turin, Italy
October 2017

Letizia Lo Presti
Salvatore Sabina

Contents

Acronyms

1D	One dimension
2D	Two dimension
3D	Three dimension
ADC	Analog-to-digital converter
AIMN	Augmentation and integrity monitoring network
AL	Alert limit
ARAIM	Advanced receiver autonomous integrity monitoring
ASP	Axle load speed profile
ATP	Automatic train protection
ATPE	Along track position error
ATPL	Along track protection level
AWGN	Additive white Gaussian noise
BG	Balise group
BR	British Rail
CAPEX	CAPital EXpenditure
CCD	Code-carrier divergence
CCS	Control-Command and Signalling
CER	Community of European Railway and Infrastructure Companies
CMC	Code minus carrier
CMD	Cold movement detection
CONUS	CONtiguous US regions
COTS	Commercial off-the-shelf
CSM	Ceiling speed monitoring
CTPL	Cross track protection level
CW	Continuous wave
DGNSS	Differential GNSS
DGPS	Differential Global Positioning System
DOP	Dilution of precision
DQM	Data quality monitoring
EBI	Emergency brake intervention

ECEF	Earth-centered earth-fixed
EEIG	European Economic Interest Grouping
EGNOS	European Geostationary Navigation Overlay System
EGNSS	European Global Navigation Satellite System
EMI	Electromagnetic interference
ENU	East North Up
EOA	End of authority
ERA	European Union Agency for Railway
ERTMS	European Rail Traffic Management System
ESA	European Space Agency
ETCS	European Train Control System
EVC	European Vital Computer
EWFA	Evil Waveform type A
EWFB	Evil Waveform type B
FAA	Federal Aviation Administration
FD	Fault detection
FDE	Fault detection and exclusion
FMEA	Fault mode and effects analysis
FS	Full supervision
FSK	Frequency shift keying
FSM	Finite state machine
FTA	Fault tree analysis
GBAS	Ground-based augmentation system
GNSS	Global Navigation Satellite System
GPS	Global Positioning System
GSA	European GNSS Agency
GSM-R	Global System for Mobile communications-Railways
HAL	Horizontal alarm limit
HMI	Hazardous misleading information
HPE	Horizontal position error
HPL	Horizontal protection level
ICAO	International Civil Aviation Organization
IF	Intermediate frequency
IM	Infrastructure manager
IMT	Integrity monitor testbed
IR	Integrity risk
ISM	Integrity support message
LAAS	Local area augmentation system
LADGNSS	Local Area Differential GNSS
LADGPS	Local area differential GPS
LED	Light-emitting diode
LGF	Local ground facility
LHCP	Left-hand circular polarized
LOA	Limit of authority
LOI	Loss of integrity

LOS	Line of sight
LRBG	Last relevant balise group
LS	Least square
MA	Movement authority
MDE	Minimum detectable ephemeris error
MEO	Medium Earth orbit
MI	Misleading information
MMSE	Minimum mean square error
MOPS	Minimum operational performance standards
MoU	Memorandum of understanding
MQM	Measurement quality monitoring
MRSP	Most restrictive speed profile
NLOS	Non Line of Sight
OBU	On-board unit
OCS	Orbit control subsystem
OPEX	OPerating EXpenditure
PDF	Probability density function
PDM	Position domain method
PE	Position error
PL	Protection level
PRC	Pseudorange correction
PVT	Position velocity and time
RAIM	Receiver autonomous integrity monitoring
RAMS	Reliability, Availability, Maintainability and Safety
RBC	Radio block center
RDM	Range domain method
R&D	Research and Development
RF	Radio frequency
RFI	Radio frequency interference
RHCP	Right-hand circular polarized
RINEX	Receiver independent exchange format
RR	Reference receiver
RS	Ranging source
RSM	Release speed monitoring
RU	Railway undertaking
S2R	Shift2Rail
SB	Stand-by
SBAS	Satellite-based augmentation system
SBI	Service brake intervention
SF	System failure
SIL	Safety integrity level
SIS	Signal in space
SNR	Signal-to-noise ratio
SoL	Safety of life
SoM	Start of mission

SQM	Signal quality monitoring
SR	Staff responsible
SS	Solution separation
SSE	Sum of the squares of the errors included in the range residuals
SSI	Solid state interlocking
SSP	Static speed profile
SV	Satellite vehicle
SvL	Supervised Location
SVN	Satellite vehicle number
SW	SoftWare
TAF	Track ahead free
TALS	Track area location server
TD	Technology demonstrator
TEC	Total electron content
TFM	Trackside functional module
THR	Tolerable hazard rate
TOA	Time of arrival
TSI	Technical specification for interoperability
TSM	Target speed monitoring
TSR	Temporary speed restriction
UNISIG	Union Industry of Signalling
VAL	Vertical alert limit
VBD	Virtual balise detector
VBG	Virtual balise group
VBR	Virtual balise reader
VBTS	Virtual balise transmission system
VDB	VHF data broadcast
VDOP	Vertical DOP
VHF	Very high frequency
VPE	Vertical position error
VPL	Vertical protection level
WAAS	Wide area augmentation system
WLS	Weighted least square

Chapter 1
Introduction and Book Objectives

Letizia Lo Presti and Salvatore Sabina

Abstract Positioning and navigation systems for applications with SoL features require the development of solutions that can ensure the integrity of the provided position information. The concept of integrity must be conjugated together with the concepts of continuity, accuracy and availability. This chapter is intended to briefly introduce these concepts, which will be dealt with more details in the following chapters. The chapter also outlines the main accuracy requirements of railway signalling systems, and the needs of innovative cost-effective railway signalling solutions and anticipates the peculiarities of the railway environment, with respect to the well-known aviation environment. Finally, this chapter summarizes the objectives of the book.

1.1 GNSS for Safety of Life Applications

GNSS consists of a constellation of satellites, which transmit signals used by a user to fix its position. Modernization of GPS and the structure of Galileo system enable SoL applications which depend not only on accurate positioning but strongly on the reliability of the position data. A main technology driver regarding SoL applications is aviation; however, maritime and railway applications can also benefit from the use of a technology similar to the one used in aviation. The key performance parameter in SoL applications is the integrity; i.e. the trust one can have in the function of the system. This means that the navigator, that is the device which computes the user position, has to guarantee the robustness of the navigation solution, to develop techniques to monitor threats and to investigate and develop algorithms to detect,

L. Lo Presti (✉)
Politecnico di Torino, c.so Duca degli Abruzzi n.24, Torino, Italy
e-mail: letizia.lopresti@polito.it

S. Sabina (✉)
Innovation and Satellite Projects, Ansaldo STS S.p.A. Via
Paolo Mantovani, 3–5, Genoa, Italy
e-mail: salvatore.sabina@ansaldo-sts.com

© Springer International Publishing AG, part of Springer Nature 2018

L. Lo Presti and S. Sabina (eds.), *GNSS for Rail Transportation*, PoliTO
Springer Series, https://doi.org/10.1007/978-3-319-79084-8_1

mitigate or exclude a faulty element of the system that could lead to unacceptable faults in the user position.

The GNSS integrity activity can be implemented autonomously by the navigator, or the user navigator can be assisted by an external *augmentation* system, able to improve the reliability of the position estimated by the navigator.

A satellite-based augmentation system (SBAS) is a system that supports wide-area or regional augmentation through the use of additional satellite broadcast messages. These messages are created thanks to the information provided by a number of ground stations, located at accurately surveyed points. These stations take measurements related to the satellite signals, which may impact the signal received by the users. The messages, created on the basis of these measurements, are sent to one or more satellites for their broadcasting to the end-users.

A ground-based augmentation system (GBAS) is a system that supports augmentation through the use of terrestrial radio messages. It is commonly composed of one or more accurately surveyed ground stations, which take measurements concerning the satellite signal, and one or more radio transmitters, which transmit the information directly to the end-user navigator. Generally, GBAS is localized to serve a geographical area within tens of kilometres (a smaller area than that covered by SBAS). The shorter the distance between the ground station that calculates the augmentation data and the assisted vehicle (end-user), the higher the accuracy of the estimated solution.

While SBASs, in principle, can be used by any kind of user (plane, ship, train, car, etc.), GBAS is normally designed for specific users (i.e. the planes landing at a specific airport), where high demanding accuracy is required [1]. Therefore, as railway applications require both high accuracy and high integrity (see Chap. 6), this book only describes GBAS, and in particular, Chap. 5 will describe the well-established GBAS augmentation system for aviation applications with the objective to give guidelines for the design of other types of GBASs. When a GBAS has to be implemented for railway applications, even if some of its elements are similar to the ones implemented in other application contexts, like aviation and maritime, this GBAS must also be either compliant with the railway standards (e.g. Cenelec EN 50126, EN 50128, EN 50129 and EN 50159) or a cross-acceptance process has to demonstrate its acceptability by the ERA.

1.2 The Concept of Integrity in Satellite Navigation

Integrity in GNSS is a quite complex concept. First and foremost, it is a guarantee for the user that the information provided by the total system, the combined ground and airborne subsystems, is correct and a critical operation can be safely accomplished. As integrity is a concept born in the aviation field, the standardized definitions and methods are mainly stated in [2].

Integrity requirements strictly depend on the application and are associated with requirements of *accuracy*, *availability* and *continuity*, which can be defined, according to [3], as given hereafter.

Accuracy. The accuracy of an estimated or measured position at a given time is the difference between the true value and the best estimate of it. Therefore, it is a measure of the degree of conformance of that position with the true position. Notice that another term, often used when dealing with an estimated position, is precision [4]. The two terms, accuracy and precision, are related to the probability density function (PDF) of the estimated position. Precision is referred to the spreading of all possible estimated values away from the mean, or the absolute known value, whereas accuracy is closeness to truth. In statistical terms, accuracy is related to the mean, while precision is related to the standard deviation. More details on this topic can be found in [4, 5].
Continuity. Continuity concerns the reliability of the position outputs of a navigation system. There is no loss of continuity when the system continues to provide navigation outputs of the specified quality for the duration of a phase of operation, assuming that the outputs were present and of specified quality at the beginning of the operation. Therefore, continuity is the ability of the total system, the combined ground and airborne subsystems, to perform its function without interruption during the intended operation. Loss of continuity can occur when a navigation system simply stops working or the augmentation system stops broadcasting signals. In other cases, continuity can be lost as a consequence of the actions of one or more integrity monitors in detecting a fault.
Integrity. Integrity is the measure of trust that can be placed in the correctness of the information supplied by the total system, the combined ground and airborne subsystems. Integrity includes the ability of the system to provide timely and valid warnings to the user (alerts) when the system should not be used for the intended operation. Integrity requirements for positioning include three elements: (1) the probability that the position error is larger than can be tolerated without annunciation, (2) the length of time (time to alert) the error can be out-of-tolerance prior to annunciation and (3) the size of the error (alert limit) that determines the out-of-tolerance condition. At signal-in-space (SIS) level, the out-of-tolerance condition is a position error that exceeds the alert limit for longer than the SIS time to alert. The true error position cannot be known [1, Appendix E].
Availability. A GNSS system is said available when the three performance parameters previously defined meet the requirements of a particular application. Availability has to be defined taking into account the ability of the system to provide a useful service within the specified coverage area. An accepted definition of availability is the long-term average probability that the accuracy, integrity and continuity requirements are simultaneously met. Therefore, the availability of a navigation system corresponds to the percentage of time that the services of the system are usable by the navigator. It depends on both the physical characteristics of the environment and the technical capabilities of the GNSS (possibly augmented) navigator. The requirements in terms of availability strongly depend on the specific applications.

A discussion on the definitions of the four performance parameters previously defined can be found in [6].

Integrity performance depends on the whole navigation system, including the space and terrestrial segments, and all the global and local augmentation elements. In this context, a key point is the design of the integrity monitoring systems: they include sophisticated algorithms employed to check the status of the sources of navigation information, not necessarily limited to GNSS, [7–12]. This book is mainly devoted to the description of these algorithms (see Chaps. 3, 4, and 5).

1.3 From Aviation to the Railway Transportation Domain

The main objective of railway signalling systems is to enable safe train movements. As a train run on a railway track, a railway signalling system must appropriately rout trains on the railway tracks and **space them** so as to avoid collisions with one another. Furthermore, due to the **high kinematic energy** associated with a moving train, the detection of events by the train driver that require the need to slow down and stop ahead does not guarantee the required safety. Therefore, at least, a signalling system must send warnings to the driver to recommend actions. There are signalling operation scenarios where a **very accurate estimation** of the train position (e.g. some metres) with a very low residual risk of failure (i.e. tolerable hazard rate (THR) equal to 10^{-9}h^{-1}) is required in all the possible environmental conditions.

The ERTMS meets these high demanding requirements. However, it is mainly used in types of lines where large investment can be applied (e.g. high-speed lines). To enforce its use on all the different types of lines (e.g. regional or low traffic lines) and to also enable the massive ERTMS implementation to achieve the expected unique European railway network, the innovation program of ERTMS [13] also foresees the use of new technologies such as GNSS positioning and other kinematics sensors IMU-based to develop new versions of cost-effective ERTMS solutions.

However, the characterization of the different types of railway lines, from regional and local lines to high-speed lines, from passenger lines to freight lines, has confirmed the hostile railway environment with respect to radio frequency interference phenomena as well as to multipath and non-line of sight conditions. This complex set of feared events leads to the not applicability of some assumptions done for the aviation application domain and of the solutions currently applied.

In addition, the key attributes of integrity, accuracy, continuity and availability defined for aviation cannot be applied as they are to the railway applications. They have to be reviewed based on the railway dependability requirements, the typical railway mission profiles and railway operational scenario.

The second part of this book will outline the peculiarity of the railway signalling systems and the related environments to understand the difference with the aviation domains and, thus, to acquire the information required for tailoring GBAS-based augmentation systems.

1.4 Objectives

This book does not claim to describe all the complex themes of GNSS-based positioning for railway applications in detail. The achievement of a standard and interoperable solution for railway applications still requires the definition of the new ERTMS system requirements and the execution of accurate performance and safety analysis agreed and accepted by the railway stakeholders. Many GSA and ESA projects have produced important and encouraging results that will be used as inputs in the context of the Shift2Rail initiative to define, develop and verify a new possible standard and interoperable solution integrated in framework of the planned ERTMS evolution. Considering the experience gained in the aviation environment, it can be expected that the enhancement process of ERTMS, based on GNSS positioning, requires years of study, experimentation and validation tests on railway pilot lines. This complex process has already been defined and started in the framework of the current on-going initiatives under the control of GSA, ESA, Shift2Rail, ERA, many national space agencies (e.g. the Italian Space Agency, ASI) and railway infrastructure managers such as Rete Ferroviaria Italiana (RFI).

Therefore, this book aims to provide support to researchers and developers who will work in this area in the coming years, providing (a) a well-structured description of augmentation systems and the integrity concept, (b) details about the GBAS augmentation system, (c) the main GNSS algorithms applied and their related properties, (d) the main basic principles of the ERTMS signalling system to safely estimate the train position, (e) the main ERTMS dependability requirements that must still be guaranteed also with the application of the GNSS technology and (f) a possible ERTMS enhancement solution also based on the GNSS positioning technology.

In particular, the first part of the book will describe the theoretical foundations of integrity and the aspect related to the GNSS domain. This first theoretical part is based on the aforementioned results of the aviation industry, and a detailed description that can be considered valid for any SoL application is also provided. It is obvious that a general approach is not possible because some solutions are strongly conditioned by the type of application. However, the methodologies described, even when they are not general, can still represent a guideline for the development of non-aviation systems that need to be developed with integrity characteristics.

The second part of this book will describe the fundamentals of the railway signalling systems and will emphasize the state-of-the-art principles currently used to estimate the train position by on-board equipment installed on train. Furthermore, Chap. 6 will also provide a detailed and rigorous description of the railway dependability requirements such as accuracy/precision, safety, schedule adherence and availability along with the tolerable hazard rate apportionment applicable to the current ERTMS system. Moreover, this chapter will also provide some examples of the peculiarities of the railway environment with respect to GNSS-related feared events. Finally, Chap. 7 will aim at describing a possible ERTMS enhancement based on the concept of virtual balises and will provide an overview of the enhanced ERTMS functional architecture, the virtual balise transmission systems and a possible THR apportionment. This chapter might be used as guidelines for evaluating the introduction of the GNSS positioning into ERTMS.

References

1. RTCA DO-245A (2004). Minimum aviation system performance standards for the local area augmentation system (LAAS)
2. RTCA/DO-208 (1991) Minimum operational performance standards for airborne supplemental navigation equipment using global positioning system (GPS), DC Radio Technical Commission for Aeronautics, Washington
3. Navipedia (2011). www.navipedia.net/index.php/GNSS_Performances
4. Hughes IG, Hase TPA (2010) Measurements and their Uncertainties A Practical Guide to Modern Error Analysis. Oxford University Press, Oxford
5. Langley RB (2010) Innovation: accuracy versus precision. GPS World
6. Pullen S (2008) What are the differences between accuracy, integrity, continuity, and availability, and how are they computed? InsideGNSS
7. Navipedia (2012). www.navipedia.net/index.php/Integrity
8. Langley RB (1999) The integrity of GPS. GPS World 10(3):6063
9. Grover Brown R (1996) Receiver autonomous integrity monitoring. In: Parkinson BW, Spilker JJ (eds) Progress In Astronautics and Aeronautics: Global Positioning System: Theory and applications, vol 2. pp 143–165 chapter 5
10. Ober PB (2003) Integrity prediction and monitoring of navigation systems. PhD thesis, Delft University Technology, The Netherlands
11. Pullen S, Walter T, Enge P (2011) Integrity for non-aviation users: Moving away from specific risk. GPS World
12. Perepetchai V (2000) Global positioning system receiver autonomous integrity monitoring. Master's thesis, School of Computer Science, McGill University, Montreal
13. ERA (2015). www.era.europa.eu/Document-Register/Documents/ERA-REP-150_ERTMS_Longer_Term_Perspective-Final_Report.pdf

Part I
GNSS Integrity

Chapter 2
Review of Common Navigation Algorithms and Measurements Errors

Letizia Lo Presti and Marco Pini

Abstract This chapter reports fundamentals of the methods of the position computation based on global navigation satellite systems. In particular, it addresses the method of position computation based on the least square method and describes the errors which affect this type of position estimate.

2.1 Methods for Position Velocity and Time (PVT) Computation

It is well known that GNSS consists of a constellation of satellites, which transmit signals used by an user to fix his position. These signals allow the users to estimate instantaneously and in real time their Position Velocity and Time (PVT) in the Earth-Centered Earth-Fixed (ECEF) reference system. The most famous GNSSs, i.e., the American Global Positioning System (GPS) and the Russian GLONASS, are based on the Time Of Arrival (TOA) concept: users determine their position evaluating the time interval between the signal transmission and the signal reception. This is possible thanks to the very accurate atomic clocks on-board the satellites, all synchronized with respect to a common time scale. It is assumed that the position of satellites is precisely known [1], whereas the time instant at which the signal leaves the satellite is embedded in the navigation message. The receiver estimates the propagation time of the signal, that in turn is multiplied by the speed of light to obtain the user-to-satellite range.

Assuming that the receiver clock is perfectly synchronized with the satellite transmitter, in the three-dimensional space, the distance d_i between the ith satellite and the user can be seen as the radius of a spheric surface having center into the position of the ith satellite, whose coordinates are represented by the vector $\mathbf{v}_i$. From the

L. Lo Presti (✉)
Politecnico di Torino, c.so Duca degli Abruzzi n.24, Torino, Italy
e-mail: letizia.lopresti@polito.it

M. Pini (✉)
Istituto Superiore Mario Boella, via P.C. Boggio n.61, Torino, Italy
e-mail: pini@ismb.it

© Springer International Publishing AG, part of Springer Nature 2018

L. Lo Presti and S. Sabina (eds.), *GNSS for Rail Transportation*, PoliTO Springer Series, https://doi.org/10.1007/978-3-319-79084-8_2

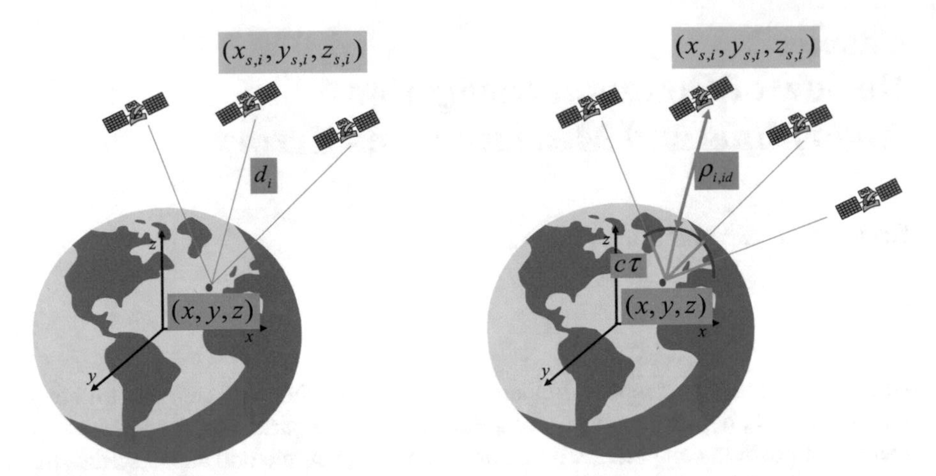

Fig. 2.1 Left: basic principle of the localization technique (TOA); right: effect of the user–receiver clock offset

intersection of at least three spheres, it is possible to compute a very precise point that represents the true user position, as depicted on the left side of Fig. 2.1. Giving the definition of distance, such as $d_i = \|\mathbf{v}_i - \mathbf{v}_U\|$, and setting a system of three equations, it is possible to obtain the solution of $\mathbf{v}_U$ representing the user coordinates in a cartesian system.

In a more realistic situation, the receiver clock is not perfectly synchronized with the transmitting satellites, therefore, the measure of the distances suffers of an unknown and common error, $c\tau$, as sketched on the right side of Fig. 2.1. This is called pseudorange, and it is defined as the sum of the true receiver–user distance and a correction factor. Analytically the theoretical pseudorange can be written as

$$\rho_{i,id} = \sqrt{(x_{s,i} - x)^2 + (y_{s,i} - y)^2 + (z_{s,i} - z)^2} + c\tau \tag{2.1}$$

while the measured pseudorange is

$$\rho_i = \rho_{i,id} + \varepsilon_i = \sqrt{(x_{s,i} - x)^2 + (y_{s,i} - y)^2 + (z_{s,i} - z)^2} + c\tau + \varepsilon_i \tag{2.2}$$

where:

- $\rho_{i,id}$ is the ideal theoretical pseudorange;
- ρ_i is the measured pseudorange;
- $(x_{s,i}, y_{s,i}, z_{s,i})$ represents the coordinates of the ith satellite position;
- (x, y, z) are the unknown coordinates of the user's position;
- τ takes into account the synchronization error between the user's clock and the satellite reference time scale;
- ε_i is a generic error affecting the measure of the pseudorange;
- c is the speed of light.

Equation (2.2) contains four unknowns, that are the user's coordinates (x, y, z) and the synchronization error τ. Therefore, a system of at least four equations is necessary to solve the problem. For simplicity purposes, we introduce a new variable $s = c\tau$ and we will analyze the estimation of the four unknowns (x, y, z, s) instead of (x, y, z, τ). Given an initial estimate of the user position (i.e., (x_0, y_0, z_0)) and an initial estimate of the clock synchronization error in meters $s_0 = c\tau_0$ the part of Eq. (2.2) not affected by errors, that is (2.1), can be approximated through the Taylor series:

$$\rho_{i,id}(x, y, z, s) = \rho_{i,id}(x_0, y_0, z_0, s_0) + \\ + (x - x_0)m_{x,i} + (y - y_0)m_{y,i} + (z - z_0)m_{z,i} + (s - s_0)m_s \tag{2.3}$$

where

$$\begin{aligned} m_{x,i} &= \left.\frac{\partial \rho_{i,id}}{\partial x}\right|_{x=x_0} \\ m_{y,i} &= \left.\frac{\partial \rho_{i,id}}{\partial y}\right|_{y=y_0} \\ m_{z,i} &= \left.\frac{\partial \rho_{i,id}}{\partial z}\right|_{z=z_0} \\ m_s &= \left.\frac{\partial \rho_{i,id}}{\partial s}\right|_{s=s_0} \end{aligned} \tag{2.4}$$

Equation (2.3) can be written for all the N_{sat} satellites in view, and all terms $\Delta\rho_{i,id} = \rho_{i,id}(x, y, z) - \rho_{i,id}(x_0, y_0, z_0)$ can be written as a vector, where the number of elements corresponds to the number $n = N_{sat}$ of satellites[1]:

$$\boldsymbol{\Delta}_{\rho,id} = \begin{bmatrix} \Delta_{\rho_{1,id}} \\ \vdots \\ \Delta_{\rho_{n,id}} \end{bmatrix} \tag{2.5}$$

This vector can be used to write the approximate expression:

$$\begin{bmatrix} \Delta_{\rho_1} \\ \vdots \\ \Delta_{\rho_n} \end{bmatrix} \cong \begin{bmatrix} \frac{\partial \rho_1}{\partial x} & \frac{\partial \rho_1}{\partial y} & \frac{\partial \rho_1}{\partial z} & \frac{\partial \rho_1}{\partial s} \\ \vdots & \vdots & \vdots & \vdots \\ \frac{\partial \rho_n}{\partial x} & \frac{\partial \rho_n}{\partial y} & \frac{\partial \rho_n}{\partial z} & \frac{\partial \rho_n}{\partial s} \end{bmatrix} \begin{bmatrix} \Delta x \\ \Delta y \\ \Delta z \\ \Delta s \end{bmatrix} + \begin{bmatrix} \varepsilon_1 \\ \vdots \\ \varepsilon_n \end{bmatrix} \tag{2.6}$$

where the pseudoranges are measured quantities, $\Delta x = x - x_0$, $\Delta y = y - y_0$, $\Delta z = z - z_0$, and $\Delta s = s - s_0 = c(\tau - \tau_0)$ and the partial derivatives are solved for $x = x_0$, $y = y_0$, $z = z_0$, and $s = s_0 = c\tau_0$, respectively.

[1] To simplify the notations in some equations, the symbol n is used to indicate the number N_{sat} of satellites in view.

Equation (2.6) can be written in a matrix form as:

$$\boldsymbol{\Delta}_\rho = \mathbf{H}\boldsymbol{\Delta}\mathbf{x} + \boldsymbol{\varepsilon} \tag{2.7}$$

where each symbol in (2.7) is immediately obtained from (2.6). Then, after computing the derivatives of (2.2) defined in (2.4), we get a matrix $\mathbf{H}$ equal to:

$$\mathbf{H} = \begin{bmatrix} \frac{x_0 - x_{s,1}}{\rho_1} & \frac{y_0 - y_{s,1}}{\rho_1} & \frac{z_0 - z_{s,1}}{\rho_1} & 1 \\ \vdots & \vdots & \vdots & \vdots \\ \frac{x_0 - x_{s,n}}{\rho_n} & \frac{y_0 - y_{s,n}}{\rho_n} & \frac{z_0 - z_{s,n}}{\rho_n} & 1 \end{bmatrix} \tag{2.8}$$

This is known as Jacobian matrix, which is the matrix of all first-order partial derivatives of a generic vector valued function F (in our case represented by (2.2), that represents a function which takes as input a real set of n elements (in our case $\boldsymbol{\Delta}_\rho$) and produces as output a real set of m elements (in our case $\boldsymbol{\Delta}\mathbf{x}$). If the function can be differentiated at a given value p, then the Jacobian matrix defines a linear map, which represents the best linear approximation of the function F, near the point p. In order to calculate a valid PVT solution, we have to solve (2.7). Since the system of equations is overdetermined and the measurements are affected by errors only an approximated solution $\hat{\boldsymbol{\Delta}}\mathbf{x}$ can be found.

2.1.1 Least Square Solution

The so-called Least square (LS) solution is obtained by introducing the error quantity:

$$\mathbf{e} = \boldsymbol{\Delta}_\rho - \mathbf{H}\hat{\boldsymbol{\Delta}}\mathbf{x} \tag{2.9}$$

where $\hat{\boldsymbol{\Delta}}\mathbf{x}$ is the approximated solution, which is obtained by performing a minimization process of $\mathbf{e}^T\mathbf{e}$ with respect to $\hat{\boldsymbol{\Delta}}\mathbf{x}$. It can be proved that this minimization leads to the equation

$$\mathbf{H}^T\mathbf{e} = \mathbf{0} \tag{2.10}$$

which means that the error vector $\mathbf{e}$ is orthogonal to the columns of $\mathbf{H}$. From (2.10), by substituting (2.9), it is easily found that the minimum-error solution is

$$\hat{\boldsymbol{\Delta}}\mathbf{x} = (\mathbf{H}^T\mathbf{H})^{-1}\mathbf{H}^T\boldsymbol{\Delta}_\rho \tag{2.11}$$

which can be rewritten in the form

$$\hat{\boldsymbol{\Delta}}\mathbf{x} = \mathbf{G}^T\boldsymbol{\Delta}_\rho \tag{2.12}$$

by introducing the notation $\mathbf{G}^T = (\mathbf{H}^T\mathbf{H})^{-1}\mathbf{H}^T$. This represents an estimate of the true value of $\boldsymbol{\Delta}\mathbf{x}$, which can be derived from (2.7), by writing

$$\mathbf{H}\boldsymbol{\Delta}\mathbf{x} = \boldsymbol{\Delta}_\rho - \mathbf{v} = \boldsymbol{\Delta}_{\rho,id} \tag{2.13}$$

where $\boldsymbol{\Delta}_{\rho,id}$ is the vector of the delta pseudoranges in the ideal case of no measurements errors.

Note that (2.12) represents also the relationship between the pseudorange errors and the position and $c\tau$ errors. Moreover the two Eqs. (2.10) and (2.13) can be associated with the concepts of range space and null space of a matrix, as it is explained in the next subsections.

2.1.2 Geometric Interpretation of Least Square Solution Using Orthogonal Subspaces

The purpose of this section is to give the general picture of the least square solution from a geometrical point of view. This interpretation will allow the reader to better understand the topic of residual-based integrity monitoring introduced in Chap. 3.

2.1.2.1 Range and Null Space Decomposition

The first step for understanding our statements is to review some results of linear algebra. We first recall the definition of the column range of $\mathbf{H}$, $Range\{\mathbf{H}\}$:

$$Range\{\mathbf{H}\} = \{\mathbf{p} \in \mathbb{R}^{N_{sat}} |\ \ \exists \mathbf{t} \in \mathbb{R}^4 : \mathbf{p} = \mathbf{H}\mathbf{t}\} \tag{2.14}$$

and the definition of the null space of $\mathbf{H}^T$, $Null\{\mathbf{H}^T\}$:

$$Null\{\mathbf{H}^T\} = \{\mathbf{p} \in \mathbb{R}^{N_{sat}} |\ \ \mathbf{H}^T\mathbf{p} = \mathbf{0}\} \tag{2.15}$$

The two spaces are completely described by the projection matrices $\mathbf{P}$ and $\mathbf{I} - \mathbf{P}$, respectively, where $\mathbf{P} = \mathbf{H}(\mathbf{H}^T\mathbf{H})^{-1}\mathbf{H}^T$. In fact, it is possible to show [2] that an equivalent definition for (2.14) and (2.15) is respectively

$$Range\{\mathbf{H}\} = \{\mathbf{p} \in \mathbb{R}^{N_{sat}} |\ \ \exists \mathbf{f} \in \mathbb{R}^{N_{sat}} : \mathbf{p} = \mathbf{P}\mathbf{f}\} \tag{2.16}$$

and

$$Null\{\mathbf{H}^T\} = \{\mathbf{p} \in \mathbb{R}^{N_{sat}} | \exists \mathbf{f} \in \mathbb{R}^{N_{sat}} : \mathbf{p} = (\mathbf{I} - \mathbf{P})\mathbf{f}\} \tag{2.17}$$

This corresponds to the decomposition of the space $\mathbb{R}^{N_{sat}}$ into two complementary orthogonal subspaces, that is

$$\mathbb{R}^{N_{sat}} = Range\{\mathbf{H}\} \oplus Null\{\mathbf{H}^T\} \tag{2.18}$$

Moreover, the following properties hold

$$\begin{cases} dim(Range\{\mathbf{H}\}) = 4 \quad (i) \\ dim(Null\{\mathbf{H}^T\}) = N_{sat} - 4 \quad (ii) \\ Range\{\mathbf{H}\} \perp Null\{\mathbf{H}^T\} \quad (iii) \end{cases} \tag{2.19}$$

Proof of (2.19),

(i) The matrix $\mathbf{H}$ is full column rank and thus a basis for the range of $\mathbf{H}$ is represented by the 4 columns of $\mathbf{H}$. Thus $dim(Range\{\mathbf{H}\}) = 4$.
(ii) Define for simplicity $\mathbf{A} = (\mathbf{I} - \mathbf{H}(\mathbf{H}^T\mathbf{H})^{-1}\mathbf{H}^T) = (\mathbf{I} - \mathbf{P})$.

- $\mathbf{A}$ is symmetric, i.e: $\mathbf{A} = \mathbf{A}^T$
- $\mathbf{A}$ is idempotent, i.e: $\mathbf{A}^2 = \mathbf{A}$
- thanks to symmetry and idempotent property $\mathbf{A}^T\mathbf{A} = \mathbf{A}^2 = \mathbf{A}$
- being $\mathbf{A}^T\mathbf{A}$ idempotent, its eigenvalues are either equal to 1 or 0
- the rank of $\mathbf{A}^T\mathbf{A} = \mathbf{A}$ is equal to the trace of $\mathbf{A}$, due to the fact that its nonzero eigenvalues are either equal to 1 or 0
- the trace of $\mathbf{A}$ is equal to

$$\begin{aligned} &\mathrm{tr}((\mathbf{I} - \mathbf{H}(\mathbf{H}^T\mathbf{H})^{-1}\mathbf{H}^T)) = \\ &\mathrm{tr}(\mathbf{I}) - \mathrm{tr}(\mathbf{H}(\mathbf{H}^T\mathbf{H})^{-1}\mathbf{H}^T) = \\ &\mathrm{tr}(\mathbf{I}) - \mathrm{tr}(\mathbf{H}^T\mathbf{H})^{-1}\mathbf{H}^T\mathbf{H}) = \\ &N_{sat} - 4 \end{aligned}$$

and thus, the matrix $(\mathbf{I} - \mathbf{H}(\mathbf{H}^T\mathbf{H})^{-1}\mathbf{H}^T)$ has rank equal to $N_{sat} - 4$ and its non zero eigenvalues are equal to 1.

(iii) The orthogonality is checked directly by (2.16), (2.17), in fact $\forall \mathbf{f}_1, \mathbf{f}_2 \in \mathbb{R}^{N_{sat}}$

$$\begin{aligned} &\mathbf{f}_2^T(\mathbf{I} - \mathbf{P})^T\mathbf{P}\mathbf{f}_1 = \mathbf{f}_2^T(\mathbf{I} - \mathbf{P})\mathbf{P}\mathbf{f}_1 = \\ &\mathbf{f}_2^T(\mathbf{P} - \mathbf{P}^2)\mathbf{f}_1 = \mathbf{f}_2^T(\mathbf{P} - \mathbf{P})\mathbf{f}_1 = 0 \end{aligned}$$

□

2.1.2.2 Least Square and Subspaces Projection

We can now use the concept introduced in Sect. 2.1.2.1 for understanding the relationship between the Least square estimation method, the subspaces projections of the measured vector of pseudoranges $\boldsymbol{\Delta}_\rho$, and the effect of the error vector $\boldsymbol{\varepsilon}$, introduced in (2.7), into the estimated user position error.

As stated in Sect. 2.1.1, the LS estimated position $\hat{\boldsymbol{\Delta}}\mathbf{x}$ is derived in such a way that the square norm of the vector $\mathbf{e}$ is minimized, where the following relationship holds

$$\mathbf{H}\hat{\boldsymbol{\Delta}}\mathbf{x} = \boldsymbol{\Delta}_\rho - \mathbf{e} \tag{2.20}$$

The problem is equivalent to find a vector $\hat{\boldsymbol{\Delta}}_\rho = \mathbf{H}\hat{\boldsymbol{\Delta}}\mathbf{x}$ such that the square norm of the vector $\mathbf{e}$ is minimized, where obviously, from (2.20),

$$\mathbf{e} = \boldsymbol{\Delta}_\rho - \hat{\boldsymbol{\Delta}}_\rho \tag{2.21}$$

It is possible to show [2] that the solution to the proposed optimization problem is

$$\hat{\boldsymbol{\Delta}}_\rho = \mathbf{P}\boldsymbol{\Delta}_\rho \tag{2.22}$$

where $\mathbf{P} = \mathbf{H}(\mathbf{H}^T\mathbf{H})^{-1}\mathbf{H}^T$ is the projection matrix introduced in Sect. 2.1.2.1. The intuitive explanation of (2.22) is the following: we want to find a vector $\hat{\boldsymbol{\Delta}}_\rho$ belonging to the range of $\mathbf{H}$ such that the distance between $\hat{\boldsymbol{\Delta}}_\rho$ and $\boldsymbol{\Delta}_\rho$ is minimized, then it can be easily understood that the optimal solution is to choose the projection of $\boldsymbol{\Delta}_\rho$ into the subspace delimited by the range of $\mathbf{H}$, from which (2.22).

Proof of (2.22), Since *Range*$\{\mathbf{H}\}$ and *Null*$\{\mathbf{H}^T\}$ form a complete subspace decomposition of $\mathbb{R}^{N_{sat}}$ we can without loss of generality write

$$\boldsymbol{\Delta}_\rho = \boldsymbol{\Delta}_{\rho_H} + \boldsymbol{\Delta}_{\rho_\perp}$$

where $\boldsymbol{\Delta}_{\rho_H}$ is the component belonging to *Range*$\{\mathbf{H}\}$ and $\boldsymbol{\Delta}_{\rho_\perp}$ the component belonging to *Null*$\{\mathbf{H}^T\}$.

The objective is to find $\hat{\boldsymbol{\Delta}}_\rho$ s.t. $||\boldsymbol{\Delta}_\rho - \hat{\boldsymbol{\Delta}}_\rho||^2$ is minimized. We can write

$$||\boldsymbol{\Delta}_\rho - \hat{\boldsymbol{\Delta}}_\rho||^2 = ||\boldsymbol{\Delta}_{\rho_H} + \boldsymbol{\Delta}_{\rho_\perp} - \hat{\boldsymbol{\Delta}}_\rho||^2$$

that can be expanded as

$$||\boldsymbol{\Delta}_{\rho_H} - \hat{\boldsymbol{\Delta}}_\rho||^2 + ||\boldsymbol{\Delta}_{\rho_\perp}||^2 - 2(\boldsymbol{\Delta}_{\rho_\perp}^T(\boldsymbol{\Delta}_{\rho_H} - \hat{\boldsymbol{\Delta}}_\rho))$$

since $(\boldsymbol{\Delta}_{\rho\perp}^{T}(\boldsymbol{\Delta}_{\rho H} - \hat{\boldsymbol{\Delta}}_{\rho})) = 0$ the solution that minimize

$$|\boldsymbol{\Delta}_{\rho H} - \hat{\boldsymbol{\Delta}}_{\rho}||^2 + ||\boldsymbol{\Delta}_{\rho\perp}||^2$$

is

$$\hat{\boldsymbol{\Delta}}_{\rho} = \boldsymbol{\Delta}_{\rho H}$$

□

It is then evident that we can decompose the error vector $\boldsymbol{\varepsilon}$, introduced in (2.7), into its component belonging to the range of $\mathbf{H}$, $\boldsymbol{\varepsilon}_H$, and its component belonging to the null space of $\mathbf{H}^T$, $\boldsymbol{\varepsilon}_{\perp}$

$$\boldsymbol{\varepsilon} = \boldsymbol{\varepsilon}_H + \boldsymbol{\varepsilon}_{\perp} \tag{2.23}$$

where the two components can be derived using the projection matrices $\mathbf{P}$ and $\mathbf{I} - \mathbf{P}$

$$\begin{cases} \boldsymbol{\varepsilon}_H = \mathbf{P}\boldsymbol{\varepsilon} \\ \boldsymbol{\varepsilon}_{\perp} = (\mathbf{I} - \mathbf{P})\boldsymbol{\varepsilon} \end{cases} \tag{2.24}$$

The component $\boldsymbol{\varepsilon}_H$ is undetectable and is the component that determines the position estimation error, while the component $\boldsymbol{\varepsilon}_{\perp}$ is the one that we can observe and on which statistical tests can be performed. The processing of this observable quantity is dealt with in Chap. 3.

The estimated position, given in (2.11), can now be written as

$$\hat{\boldsymbol{\Delta}\mathbf{x}} = (\mathbf{H}^T\mathbf{H})^{-1}\mathbf{H}^T\boldsymbol{\Delta}_{\rho} = \boldsymbol{\Delta}\mathbf{x} + (\mathbf{H}^T\mathbf{H})^{-1}\mathbf{H}^T\boldsymbol{\varepsilon} \tag{2.25}$$

from which we can derive the estimated position error

$$\hat{\boldsymbol{\delta}\mathbf{x}} = (\mathbf{H}^T\mathbf{H})^{-1}\mathbf{H}^T(\boldsymbol{\varepsilon}_H + \boldsymbol{\varepsilon}_{\perp}) = (\mathbf{H}^T\mathbf{H})^{-1}\mathbf{H}^T(\boldsymbol{\varepsilon}_H) \tag{2.26}$$

since the component $\boldsymbol{\varepsilon}_{\perp}$ belongs to the null space of $\mathbf{H}^T$. It is clear that the component of $\boldsymbol{\varepsilon}$ that generates a position estimation error is only $\boldsymbol{\varepsilon}_H$. On the other hand, the component that we can observe is

$$(\mathbf{I} - \mathbf{H}(\mathbf{H}^T\mathbf{H})^{-1}\mathbf{H}^T)\boldsymbol{\varepsilon} = \boldsymbol{\varepsilon}_{\perp} \tag{2.27}$$

These notions will hopefully help the reader in understanding the concepts introduced in Chap. 3.

2.1.3 Weighted Least Square Solution and Subspace Projection

The LS solution described in Sect. 2.1.1 can be modified by including a matrix of weights $\mathbf{W_e}$ to be given to each element of the vector $\mathbf{e}$ introduced in (2.9). In this case, the solution of the system becomes

$$\hat{\boldsymbol{\Delta}}\mathbf{x} = (\mathbf{H}^T\mathbf{W}\mathbf{H})^{-1}\mathbf{H}^T\mathbf{W}\boldsymbol{\Delta}_\rho \tag{2.28}$$

where $\mathbf{W} = \mathbf{W}_e^T\mathbf{W}_e$. A common criterion is to set the weights according to the C/N_0 ratio, which is typically provided by any GNSS receiver. In this way, the pseudoranges with lower C/N_0 will have a lower impact in the estimation of the position.

In a practical case, if we know that the variance of the measurement errors is different for different pseudoranges, i.e., the covariance $\boldsymbol{\Sigma}$ of the error vector $\boldsymbol{\varepsilon}$ is not a scaled identity matrix, and

$$\boldsymbol{\varepsilon} \sim N(\mathbf{0}, \boldsymbol{\Sigma}) \tag{2.29}$$

the optimal solution in a minimum mean square error (MMSE) sense is a Weighted Least square (WLS) estimation algorithm. It can be easily shown that if we define a new measurement vector $\boldsymbol{\Delta}_\rho^*$ as

$$\boldsymbol{\Delta}_\rho^* = \boldsymbol{\Sigma}^{-\frac{1}{2}}\boldsymbol{\Delta}_\rho \tag{2.30}$$

and we define a new geometry matrix as

$$\boldsymbol{H}^* = \boldsymbol{\Sigma}^{-\frac{1}{2}}\boldsymbol{H} \tag{2.31}$$

then we can estimate the unknown user position in an optimal way as

$$\hat{\boldsymbol{\Delta}}\mathbf{x} = (\mathbf{H}^{*T}\mathbf{H}^*)^{-1}\mathbf{H}^{*T}\boldsymbol{\Delta}_\rho^* \tag{2.32}$$

that is perfectly equivalent to (2.28) if we choose $\mathbf{W} = \boldsymbol{\Sigma}^{-1}$. Also in this case, we can interpret the problem in terms of finding the vector $\hat{\boldsymbol{\Delta}}_\rho^*$ such that the square norm of

$$\mathbf{e}^* = \boldsymbol{\Delta}_\rho^* - \hat{\boldsymbol{\Delta}}_\rho^* \tag{2.33}$$

is minimized, where $\hat{\boldsymbol{\Delta}}_\rho^* = \mathbf{H}^*\hat{\boldsymbol{\Delta}}\mathbf{x}$. The solution to this problem is, from (2.32)

$$\hat{\boldsymbol{\Delta}}_\rho^* = \mathbf{H}^*(\mathbf{H}^{*T}\mathbf{H}^*)^{-1}\mathbf{H}^{*T}\boldsymbol{\Delta}_\rho^* \tag{2.34}$$

Equation (2.34) gives us the possibility of interpreting again the WLS algorithm in terms of projection into subspaces: we start with a measurement vector $\boldsymbol{\Delta}_\rho$ and we transform it into a new vector $\boldsymbol{\Delta}^*_\rho$ trough the whitening transformation of (2.30), then we obtain the solution to the optimization problem by projecting the vector $\boldsymbol{\Delta}^*_\rho$ into the range of $\mathbf{H}^*$. In fact, we can rewrite (2.34) as

$$\hat{\boldsymbol{\Delta}}^*_\rho = \mathbf{P}^* \boldsymbol{\Delta}^*_\rho \tag{2.35}$$

where $\mathbf{P}^* = \mathbf{H}^*(\mathbf{H}^{*T}\mathbf{H}^*)^{-1}\mathbf{H}^{*T}$ is the matrix that projects into the range of $\mathbf{H}^*$.

2.1.4 *Covariance Matrix*

It can be proven that the covariance matrix of $\hat{\boldsymbol{\Delta}}\mathbf{x}$, defined as $\mathbf{C}_{\hat{\boldsymbol{\Delta}}\mathbf{x}} = E\{\hat{\boldsymbol{\Delta}}\mathbf{x}\hat{\boldsymbol{\Delta}}\mathbf{x}^T\}$, can be written according to the following two statements if a weight matrix is present or not. In the case without weights, we have:

$$\mathbf{C}_{\hat{\boldsymbol{\Delta}}\mathbf{x}} = (\mathbf{H}^T\mathbf{H})^{-1}\mathbf{H}^T\mathbf{C}_{\Delta_\rho}\mathbf{H}(\mathbf{H}^T\mathbf{H})^{-1} \tag{2.36}$$

On the contrary, when the weights are used, (2.36) becomes

$$\mathbf{C}_{\hat{\boldsymbol{\Delta}}\mathbf{x}} = (\mathbf{H}^T\mathbf{W}\mathbf{H})^{-1}\mathbf{H}^T\mathbf{W}\mathbf{C}_{\Delta_\rho}\mathbf{W}\mathbf{H}(\mathbf{H}^T\mathbf{W}\mathbf{H})^{-1} \tag{2.37}$$

where $\mathbf{C}_{\Delta_\rho}$ is the covariance matrix of the available measurements. The estimation variance is a fundamental parameter to have an idea about the quality of the estimates. As we can see, it depends on the accuracy of the pseudorange measurements (i.e., $\mathbf{C}_{\Delta_\rho}$ matrix) and on the matrix $\mathbf{H}$.

2.1.5 *Errors Due to the Receiving Hardware and Local Environment*

2.1.5.1 Multipath

Multipath represents a tremendous error source, difficult to mitigate, mainly in those applications where the antenna cannot be sited away from reflecting obstacles. From a general perspective, multipath is defined as the presence of multiple signal paths between a transmitter and a receiver due to reflections and diffractions. For GNSS receivers, the only desired signal is the one that propagates following the Line Of Sight (LOS). In fact, as explained above, in order to determine the user position the receiver evaluates the time taken by the signal to travel from the satellites to the receiving antenna. It is evident that if there are signal replicas superimposed to the LOS, these induce signal distortions and might prevent the digital tracking loops to finely track the code and carrier phase. It is quite difficult to mitigate the presence of multipath at the front-end level. If the multipath is not blocked at the antenna, it cannot be cut off neither at the Radio Frequency (RF) and at the Intermediate Frequency (IF) stage of the receiver front end and propagates down to the ADC, affecting the digital signal processing. The choice of Right Hand Circularly Polarized (RHCP) signals in satellite navigation was mainly driven by the possibility of mitigating the effect of multipath at the antenna. Ideally, a single reflection of the LOS flips the polarization of the electric field, which becomes Left Hand Circularly Polarized (LHCP). In this way, the antenna strongly reduces the effect of multipath coming from a single reflection. However, an even number of reflections restores the original polarization and in this case the antenna cannot mitigate the multipath effect.

Considering a real scenario, characterized by many obstacles near to the receiving antenna, multipath can be the main source of positioning error. It is worth remarking that in addition to the amplitude, delay and phase, the multipath rays can have different frequencies with respect to the LOS. Introducing f_0 as the carrier frequency, the ith multipath can be represented as:

$$m_i(t) = |a_i(t)|\, s(t - \tau_i) \cos(2\pi f_0 t + \angle a_i(t)) \tag{2.38}$$

where $|a_i(t)|$ models time variant attenuation, $\angle a_i(t)$ takes into account phase rotations due to reflections and refractions, τ_i is the delay with respect to the LOS. The instantaneous carrier frequency of the multipath depends on $\angle a_i(t)$, in fact:

$$f_i(t) = f_0 + \frac{1}{2\pi} \cdot \frac{d\angle a_i(t)}{dt} \tag{2.39}$$

This phenomenon depends on the non-stationary scenario and on the user motion. Furthermore, note that if the LOS is not shadowed, then a single reflection can be reasonably assumed weaker than the LOS signal [3]. Roughly speaking, when the multipath has a delay shorter than the chip of the spreading code, the discrimination function used by digital code tracking loops is distorted. In turn, this produces a misalignment between the incoming spreading code and a local replica generated internally, that the receiver uses to estimate the pseudorange. In other words, such local replica is not perfectly synchronized to the incoming code, but is ahead or behind the incoming direct-path signal, depending upon the multipath delay. This yields to a systematic pseudorange error.

Multipath is a well-known error source in satellite navigation and a variety of countermeasures have been studied and proposed for different applications and types of receivers, since several years. Interested readers can address their curiosity and find a number of references in a wide scientific literature.

2.1.5.2 External Interfering Signals

The monitoring of Radio Frequency Interference (RFI) has a fundamental role in many applications, especially when a high level of accuracy and continuity is required. Indeed, several electromagnetic sources might degrade GNSS measurements and can be either in band disturbances (i.e., secondary harmonics due to nonlinearity distortions, generated in the transmission of other communication systems, intentional jamming) or out-of-band (i.e., strong signals that occupy frequency bands close to GNSS bands).

As mentioned above, one of the main characteristics of GNSS signals is the low level of signal power reaching the receiving antenna. Despite of the weakness of the signals, thanks to the spread spectrum nature of the transmission, navigation receivers recover timing information and compute the user's position, exploiting the gain obtained at the output of the correlation block [4]. Among all the different error sources that corrupt satellite navigation waveforms, the presence of interference is particularly harmful, since in some cases their effect cannot be mitigated by the correlation process. In fact, even if the spread spectrum technique is theoretically able to mitigate the presence of jammers in the bandwidth of interest, the actual limitation is the dynamic of the receiver front-end hardware. The presence of interference on the input signal affects the regular receiver signal processing in different ways and relates to the nature of the interferers (i.e., power, bandwidth) and to the type of receiver front ends. Depending on how much signal power passes through the first stages of the receiver, interferers affect the carrier and the code tracking loops, which results in deteriorized GPS observables or in a complete loss of lock in severe cases [5]. If the interfering signal is sufficiently strong, the receiver can be blinded and forced to stop the signal processing, with a clear risk for the service continuity. The relevance of the degradation factor depends on the position of the receiving GNSS antenna with respect to the undesired interfering sources. The number of electromagnetic sources that are candidate to become interferers for GNSS signals is large and increases with the advent of new wireless systems and with the integration of GNSS with other technologies. All of these sources have a different signal structure and are often classified with respect to their characteristics in time and frequency domains (e.g., Continuous Wave (CW), pulsed signals, chirp).

In addition to not structured interfering signals, in the last years a growing concen is represented by *meaconing* and *spoofing*, that can be classified as intentional interference. The first refers to the rebroadcasting of delayed GNSS signals without any distinction between the Signal In Space (SIS) from different satellites. The second is the generation and transmission of false GNSS signals, with the intent to produce erroneous positions within the victim receiver without disrupting the GNSS operations. Meaconing is a rather simple attack to accomplish and induces the victim to provide PVT with degraded accuracy. On the other hand, spoofing is more malicious and can fool the receiver without any notice. Both meaconing and spoofing are tremendous threats for the GNSS integrity. Especially over the last decade, a variety of countermeasures have been studied and proposed against meaconing and spoofing. More details can be found in recent scientific literature, in particular in [6] that provides an overview of the problem, recalling the most updated countermeasures.

2.1.5.3 Poor Receiver Clock Stability

Another distortion of the received signal is originated locally within the receiver. The analog signal processing involves filtering, amplification, and frequency downconversion. Generally, front ends of GNSS receivers follow a heterodyne architecture: the received Radio Frequency (RF) signal is mixed with a tone generated by a local oscillator, to have one component of the resulting product centered to a desired lower frequency, namely the Intermediate Frequency. Neglecting for simplicity the distortions due to nonlinearities of amplifiers and filters, as well as the thermal noise, the GNSS signal at RF entering into the mixed can be modeled as:

$$s_{RF}(t) = G_1\sqrt{2P_R}c(t-\tau)d(t-\tau)\cos(2\pi(f_R F + f_d)t + \phi_{RF}(t)) \tag{2.40}$$

where:

- G_1 represents the gain introduced by the RF stage, whereas P_R is received signal power;
- $c(t)$ and $d(t)$ are the spreading code and the navigation data bits transmitted by the satellite and received after τ s, that is the propagation delay;
- $\cos(2\pi(f_R F + f_d)t + \phi_{RF})$ is the signal carrier at the central frequency f_{RF}, that is shifted the term f_d, due to the Doppler effect;
- ϕ_{RF} is an unknown phase.

The tone generated by the local oscillator can be written as:

$$s_{LO}(t) = cos(2\pi f_{LO}t + \phi_{LO}(t)) \tag{2.41}$$

where f_{LO} and $\phi_{LO}(t)$ are the frequency and the phase of the local signal. In Eq. (2.41), the phase term depends on time to indicate that the local oscillator is not an ideal component and can have frequency instability. In fact, real components do not generate a pure tone centered always on the same frequency, but their instantaneous frequency is not constant and varies with time.

The mixer outputs the product of the two signals. The component beating at $f_{RF} + f_d + f_{LO}$ is filtered by the Intermediate Frequency (IF) filter following the mixer, while the other component is:

$$s_{IF}(t) = \frac{G_1\sqrt{2P_R}}{2}c(t-\tau)d(t-\tau)\cdot\cos(2\pi f_{IF} + f_d t + \phi_{RF} - \phi LO(t)) \tag{2.42}$$

where $f_{IF} = f_{RF} - f_{LO}$. Taking the derivative of the instantaneous phase of the signal, we can compute the instantaneous frequency according to the definition, that is:

$$f_(t) = f_{IF} + fd - \frac{1}{2\pi}\frac{d\phi_{LO}(t)}{dt} \tag{2.43}$$

The last term in Eq. (2.43) represents a further shift in frequency, only due to the local oscillator. The signal is then amplified, filtered, and digitalized, but this term remains. The first stage of the digital signal processing after the Analog to Digital Converter (ADC), that is the signal acquisition, is not able to discern the Doppler and the shift due to the phase variation of the local oscillator. However, note that the term is common on all channels and can be estimated and removed by the navigation algorithms.

If a poor quality local oscillator is used in the front end, the phase variations can be fast in time. In this case, the corresponding frequency shift can be much higher than the real Doppler, producing an effect similar to that experienced by high dynamic applications, with significant accelerations between the satellite and the user's receiver. Finally note that traditional front-end architectures derive the sampling clock from the local oscillator. Therefore, the lower the phase variations of the local tone, the more constant is the sampling rate.

2.1.5.4 Thermal Noise

GNSSs employ constellations of Medium Earth Orbit (MEO) satellites. For instance, GPS satellites are into 20.200 Km circular orbits, inclined at 55° and placed in 6 different orbit planes. Using conventional hardware, the power of the received signal is extremely low, roughly speaking, 20 dB lower than the noise power. The received signal needs to be suitably conditioned by amplifiers, filters, and mixers within the GNSS front end, before being digitalized and processed by GNSS signal algorithms, that leverage on the gain obtained by the spread spectrum nature of the signal structure. The extremely low level of the received GNSS signal is certainly a unique feature in the field of radio transmission.

The presence of noise must be carefully considered in the design of GNSS signal processing algorithms, because it affects the accuracy of code and carrier measurements, that are at the basis of the pseudorange estimate. In challenging environments, when the signal-to-noise ratio is particularly low, the thermal noise is a dominant source of error, because it masks the signal characteristics that need to be be finely estimated to compute the PVT. If the received signals are not well conditioned, the thermal noise can induce the receiver to compute the PVT with poor accuracy, with the risk of not matching the application requirements. As a general rule of thumb, the measurement error due to thermal noise varies with the signal strength, which, in turn, increases with higher satellite elevation angles.

References

1. Kaplan E, Hegarty C (2006) Understanding GPS: Principles And Applications, 2nd edn. Artech House, Massachusetts
2. Dattorro J (2005) Convex Optimization & Euclidean Distance Geometry

3. Van Nee RDJ (1992) Multipath effects on GPS code phase measurements. J Inst Navig 39(2):177–190
4. Misra P, Enge P (2006) Global Positioning System. Signal, Measurements and Performance. Ganga-Jamuna Press, MA
5. Bastide F, Akos DM, Macabiau C, Roturier B (2003) Automatic gain control (AGC) as an interference assessment tool. Proceedings of ION GPS 16th technical meeting of the satellite division of the institute of navigation, Portland, Oregon
6. Dovis F (ed) (2015) GNSS Interference Threats and Countermeasures. Artech House, London

Chapter 3
Fundamentals of Integrity Monitoring

Letizia Lo Presti and Giulio Franzese

Abstract This chapter describes the methods of integrity monitoring, necessary to verify if all the satellites involved in the PVT computation are healthy or not. In particular, we will see the solution adopted by a stand-alone receiver equipped with a system able to check if the hypothesis of nominal conditions (i.e. when all the satellites are healthy) can be considered valid. This is a fundamental step before evaluating the confidence interval associated to the estimated position. The reason why integrity monitoring is a necessary step for the evaluation of the confidence interval is briefly described in Sect. 3.1, while the remainder of the chapter is devoted to the methods of fault detection (FD), and fault detection and exclusion (FDE), generally implemented in the algorithms of receiver autonomous integrity monitoring (RAIM) systems.

Note The algorithms described in this chapter are based on known results of the estimation and decision theory and often require long and complex proofs. It is advisable for the reader to read the chapter first by skipping all the proofs in order to capture the general meaning of the described methods. The proofs can be read at a later stage if it is necessary to study in depth the FD and FDE techniques.

3.1 Evaluation of the Confidence Interval in an AWGN Model

The pseudoranges are affected by different sources of error, and this inaccuracy is reflected in the user's position estimation as well. Since the positioning error is not a deterministic value, it has to be modelled as a random variable. If we could

L. Lo Presti (✉) · G. Franzese
Politecnico di Torino, c.so Duca degli Abruzzi n.24, Torino, Italy
e-mail: letizia.lopresti@polito.it

G. Franzese
e-mail: giulio.franzese@polito.it

© Springer International Publishing AG, part of Springer Nature 2018

L. Lo Presti and S. Sabina (eds.), *GNSS for Rail Transportation*, PoliTO Springer Series, https://doi.org/10.1007/978-3-319-79084-8_3

have information about the PDF of such process, we could know the probability with which the error overcomes a certain threshold. At the same way, we could fix this probability and derive the threshold. This threshold identifies a bound T_e of the position error. This bound can be computed for each position coordinate and used to affirm that the position coordinates x, y, and z are in the range

$$
\begin{aligned}
&(\hat{x} - T_{e,x}\,,\ \hat{x} + T_{e,x})\\
&(\hat{y} - T_{e,y}\,,\ \hat{y} + T_{e,y})\\
&(\hat{z} - T_{e,z}\,,\ \hat{z} + T_{e,z})
\end{aligned}
$$

where $\hat{x}$, $\hat{y}$, and $\hat{z}$ are the estimated coordinates, obtained as shown in Sect. 2.1, and $T_{e,x}$, $T_{e,y}$, and $T_{e,x}$ are the bounds of each coordinate. These values are the confidence intervals, generally used in all the measurement processes. It is evident that there exists a residual risk that the true position is outside the confidence intervals. The key idea is to accept this residual risk, by fixing this probability and to compute the confidence interval as a consequence.

In the GNSS literature, the target probability P_T used to compute the confidence interval is derived from another probability called integrity risk (IR), that will be introduced in Chap. 4. For the time being, we can ignore the definition of IR and the derivation of the target P_T from IR, and we start our analysis from P_T. The confidence interval associated to the target P_T is called protection level (PL).

In the GNSS applications, the target IR and the PL are generally associated to the vertical and horizontal (radial) coordinates in the East-North-Up (ENU) reference frame. In this case, the objective is to compute the vertical protection level (VPL) and the horizontal protection level (HPL).

The computation of PLs is not an easy task, especially for two reasons.

- The definition of the PDF of the position error is not trivial and depends on the environment.
- The derivation of PL from the target probability is generally difficult, and conservative solutions have to be generally adopted.

We will address the problem of the PL computation in the remainder of this book. We start here, in Sect. 3.1.1, with the PL computation in the very simple case of a single coordinate affected by additive white Gaussian noise (AWGN). How to compute the PL in more complex cases will be addressed in Chap. 4.

3.1.1 PL of a Single Coordinate, in Case of Zero-Mean Gaussian Position Error

Let us suppose to compute a quantity (in our case a position coordinate) that we can model as a random variable of the type $X = x_t + E$, where x_t is the true value of the computed quantity and E is an error modelled as a random variable. We introduce now the probability of the event $|E| > T_e$, that is

$$P_T = \Pr(|E| > T_e)$$

that specifies the probability to have an error greater than a generic value T_e. If the PDF $f_E(e)$ of E is an even function, this probability can be expressed as

$$P_T = 2\int_{T_e}^{\infty} f_E(e)de$$

In case of a Gaussian PDF characterized by zero-mean value and variance σ^2, this equation can be rewritten as

$$P_T = 2\int_{T_e}^{\infty} \frac{1}{\sqrt{2\pi}\sigma} e^{\frac{e^2}{2\sigma^2}} de = \frac{2}{\sqrt{\pi}} \int_{T_e/\sqrt{2}\sigma}^{\infty} e^{z^2} dz = \operatorname{erfc}\left(\frac{T_e}{\sqrt{2}\sigma}\right)$$

from which, after having fixed the probability, we can write

$$\frac{T_e}{\sqrt{2}\sigma} = \operatorname{erfcinv}(P_T)$$

As a consequence, if P_T is a target probability, the value of PL is

$$T_e = \operatorname{erfcinv}(P_T)\sqrt{2}\sigma = k_\sigma(P_T)\sigma \tag{3.1}$$

where

$$k_\sigma(P_T) = \operatorname{erfcinv}(P_T)\sqrt{2} \tag{3.2}$$

Table 3.1 shows some values of $k_\sigma(P_T)$ versus P_T.

In our case, we are interested in the confidence intervals associated to the position coordinates (x, y, z) of the centre of phase of the antenna of a GNSS receiver. First of all, we have to verify if (3.1) can be applied to evaluate the confidence intervals T_x, T_y, and T_z of the estimates of (x, y, z). This means that we have to verify if the AWGN model can be adopted for the three estimated coordinates. Since they are not directly measured, but are estimated from the measured pseudoranges, the first issue is to verify if the AWGN model is valid for the position errors induced by the errors of the measured pseudoranges. In the LS solution, the estimated coordinates $(\hat{x}, \hat{y}, \hat{z})$ are obtained from (2.11) and (2.28), which are affected by zero-mean Gaussian errors if the pseudorange errors are zero-mean Gaussian random variables. This is true if the Jacobian matrix $\mathbf{H}$ can be considered deterministic. This assumption can be reasonably accepted in practice. On the contrary, the situation is much more complex when the pseudorange errors cannot be considered zero-mean Gaussian random variables, as it will be clear in Chap. 4.

Table 3.1 Values of $k_\sigma(P_T)$ versus P_T

P_T	$k_\sigma(P_T)$
10 − 1	1.6449
10 − 2	2.5758
10 − 3	3.2905
10 − 4	3.8906
10 − 5	4.4172
10 − 6	4.8916
10 − 7	5.3267
10 − 8	5.7307
10 − 9	6.1094
10 − 10	6.4670
10 − 11	6.806

3.1.2 Verification of the AWGN Model

Before using (3.1) for the three position coordinates, it is necessary to verify if the AWGN model is valid, and this has to be done at each epoch, that is, at each new position fix. Notice that this means that we have to verify if the PDF of the estimated coordinates are Gaussian distributed, the means are zero, and the standard deviations $\sigma_{W,x}$, $\sigma_{W,y}$, and $\sigma_{W,z}$ of the errors of the three coordinates are the ones associated to the adopted model. The punctual verification of these conditions is not trivial, and some simplified approaches are proposed in the literature. A classical technique is the one adopted by the RAIM module, which tries to identify if a fault is present in each received SIS, so as to possibly exclude the satellites working in faulty condition. The inherent idea behind this approach is that the AWGN model can be considered valid if no fault is present in the computed PVT.

3.1.3 Introduction to RAIM

RAIM is a technique that uses an overdetermined solution to perform a consistency check on the satellite measurements [1]. RAIM is implemented in the receiver and is implemented through a category of algorithms that measure the integrity of the navigation solution provided by the system at hand, i.e. the trust that can be placed in the correctness of the information supplied by the navigation system, as well as at providing timely warnings to users when the system should not be used for navigation [2]. The terms *receiver autonomous* means that, in principle, RAIM does not require external support to estimate such a trust (e.g. terrestrial monitoring infrastructures, terrestrial communication channels, non-GNSS sensors,...). As affirmed in [3], the RAIM is a good technique, is quite robust, can be implemented with a low computational effort, and for these reasons, it is generally adopted in the commercial receivers.

To detect the presence of a fault in the estimated position it is necessary to introduce a test statistic to be compared against a threshold, in order to verify two possible hypotheses:

$$\mathscr{H}_0 : \text{No fault is present (Nominal condition)}$$
$$\mathscr{H}_1 : \text{A fault is present (Fault condition)}$$

In [4], Parkinson and Axelrad propose a test statistic working in the range domain and based on the definition of *range residual*. Another possible approach works in the position domain, and it is called *solution separation test*. These two methods are not equivalent. The first one is used for FD, while the second one allows also the identification of the satellites to be excluded.

Details on FD and FDE methods are given in this chapter.

3.2 Fault Detection in the Range Domain

The fault detection method working in the range domain is based on the idea to compare the measured pseudoranges with the so-called reconstructed pseudoranges, obtained starting from the computed position.

The **measured pseudorange** is based on the signal transit time from satellite to user, corrected for satellite clock errors, ionospheric and tropospheric delays, relativistic effects, satellite clock offset, group delay, etc. This pseudorange measurement contains errors due to the satellite and user clock model inaccuracies, propagation link delay model errors, multipath, interference, receiver noise, and interchannel biases. Then, the measured pseudorange can be written as:

$$\rho = d_0 + b - \varepsilon_c - \varepsilon_l - \varepsilon_m - \varepsilon_r \tag{3.3}$$

where

- d_0 is the true distance
- b is the clock offset
- ε_c is the user clock bias estimation error, i.e. uncorrected satellite clock error
- ε_l includes propagation link errors, i.e. uncorrected ionospheric and tropospheric delays
- ε_m is the multipath error
- ε_r is the receiver error

The **reconstructed range**[1] is the range between the receiver current estimate of position and the satellite position derived from the ephemeris data provided in the navigation message. This is a geometrical range, which is affected by the ephemeris errors and the position errors introduced by the adopted PVT computation algorithm. Then, the reconstructed range can be written as:

$$d = d_0 + \varepsilon_e + \varepsilon_{\mathrm{PVT}} \tag{3.4}$$

where ε_e is the ephemeris error, and $\varepsilon_{\mathrm{PVT}}$ is the error introduced by the PVT computation algorithm. Notice that ε_e is not present in (3.3), since the measured pseudorange does not use the ephemeris, but only the time of flight. In [4], the authors affirm that the only significant error in (3.4) is due to the satellite ephemeris, and then they write the reconstructed range as:

$$d = d_0 + \varepsilon_e \tag{3.5}$$

3.2.1 Range Residual Method

The range residual for the ith satellite is defined, [4], as the difference:

$$\hat{\varepsilon}_{rr,i} = d_i - (\rho_i - b) \tag{3.6}$$

where for each SV i, d_i is the reconstructed range, b is the clock offset, and $(\rho_i - b)$ is the measured range. By substituting (3.3) and (3.4) in (3.6), the range residual becomes

$$\hat{\varepsilon}_{rr,i} = \varepsilon_e + \varepsilon_{\mathrm{PVT}} + \varepsilon_c + \varepsilon_p + \varepsilon_m + \varepsilon_r \tag{3.7}$$

The sum of the squares of the errors included in the range residuals (SSE) plays the role of the basic observable in many RAIM methods. It is defined as:

$$\varepsilon_{\mathrm{SSE}} = \hat{\boldsymbol{\varepsilon}}_{rr}^T \hat{\boldsymbol{\varepsilon}}_{rr} = \mathrm{trace}(\hat{\boldsymbol{\varepsilon}}_{rr} \hat{\boldsymbol{\varepsilon}}_{rr}^T) \tag{3.8}$$

where

$$\hat{\boldsymbol{\varepsilon}}_{rr} = \begin{bmatrix} \hat{\varepsilon}_{rr,1} \\ \hat{\varepsilon}_{rr,2} \\ \vdots \\ \hat{\varepsilon}_{rr,N_{sat}} \end{bmatrix} \tag{3.9}$$

[1]In [4] this is called *predicted range*. We prefer to adopt the term reconstructed or computed range, proposed by other authors, [5], as it seems more adequate.

and N_{sat} is the number of satellites used in the PVT. The quantity ε_{SSE} can be used as test statistic to detect the presence of a fault in the estimated position. Other authors, [4], use a normalized version of the squared $\varepsilon_{textSSE}$, defined as

$$r_{\text{LS}} = \sqrt{\frac{\varepsilon_{\text{SSE}}}{N_{sat} - 4}} = \sqrt{\frac{\hat{\boldsymbol{\varepsilon}}_{rr}^T \hat{\boldsymbol{\varepsilon}}_{rr}}{N_{sat} - 4}} \tag{3.10}$$

Both test statistics are the basic observables in many RAIM methods. The reason of the normalization in (3.10) will be clear in Sect. 3.2.2.

Notice that calculating these test statistics involves some matrix manipulations, but these are not worse than implementing the PVT which is done routinely in any receiver, starting from the overdetermined system of linear equations $\boldsymbol{\Delta}_\rho = \mathbf{H}\boldsymbol{\Delta}\mathbf{x} + \boldsymbol{\varepsilon}$ given in (2.7), where the elements $\varepsilon_i, i = 1, 2, \ldots, N_{sat}$, of the vector $\boldsymbol{\varepsilon}$ are the errors of each pseudorange; therefore,

$$\boldsymbol{\varepsilon} = \begin{bmatrix} \varepsilon_1 \\ \varepsilon_2 \\ \vdots \\ \varepsilon_{N_{sat}} \end{bmatrix} \tag{3.11}$$

Equation (2.7) is written in terms of deltapseudorange and deltapositions. It is evident that these delta quantities can be used to evaluate the same range residuals introduced in this section.

Next step is to evaluate the threshold to be applied to the test statistic to decide if there is a fault in the estimated position. To set this threshold, it is necessary to know the statistical characteristics of the test statistic. For this reason, we need to find the relationship between ε_{SSE} and the pseudorange error vector $\boldsymbol{\varepsilon}$, whose statistical characteristics are assumed known.

3.2.1.1 Relationship Between Range Residual and Pseudorange Error

It is possible to prove that the two vectors $\boldsymbol{\varepsilon}$ and $\hat{\boldsymbol{\varepsilon}}_{rr}$ are related each other in the following form:

$$\hat{\boldsymbol{\varepsilon}}_{rr} = (\mathbf{I}_{N_{sat}} - \mathbf{P})\boldsymbol{\varepsilon} \tag{3.12}$$

where $\mathbf{I}_{N_{sat}}$ is the unit matrix $N_{sat} \times N_{sat}$, $\mathbf{P} = \mathbf{H}(\mathbf{H}^T\mathbf{H})^{-1}\mathbf{H}^T$ is the projection matrix introduced in Sect. 2.1.2.1, and $\mathbf{H}$ is the Jacobian matrix introduced in (2.8).

Proof of (3.12), [6]. Let us start with the expressions of the delta pseudorange vector $\boldsymbol{\Delta}_\rho = \mathbf{H}\boldsymbol{\Delta}\mathbf{x} + \boldsymbol{\varepsilon}$, and the vector of the estimated incremental positions $\hat{\boldsymbol{\Delta}}\mathbf{x} = (\mathbf{H}^T\mathbf{H})^{-1}\mathbf{H}^T\boldsymbol{\Delta}_\rho$ given in (2.11), from which the vector of the reconstructed pseudorange can be written as:

$$\hat{\boldsymbol{\Delta}}_\rho = \mathbf{H}\hat{\boldsymbol{\Delta}}\mathbf{x} = \mathbf{H}(\mathbf{H}^T\mathbf{H})^{-1}\mathbf{H}^T\boldsymbol{\Delta}_\rho = \mathbf{P}(\mathbf{H}\boldsymbol{\Delta}\mathbf{x} + \boldsymbol{\varepsilon}) \tag{3.13}$$

where $\mathbf{P} = \mathbf{H}(\mathbf{H}^T\mathbf{H})^{-1}\mathbf{H}^T$ has been just introduced in (3.12). Now it is possible to write the vector of the range residuals as

$$\hat{\boldsymbol{\varepsilon}}_{rr} = \boldsymbol{\Delta}_\rho - \hat{\boldsymbol{\Delta}}_\rho = (\mathbf{I}_{N_{sat}} - \mathbf{P})\boldsymbol{\Delta}_\rho = (\mathbf{I}_{N_{sat}} - \mathbf{P})\mathbf{H}\boldsymbol{\Delta}\mathbf{x} + (\mathbf{I}_{N_{sat}} - \mathbf{P})\boldsymbol{\varepsilon}$$

where the first term is zero, since

$$\begin{aligned}(\mathbf{I}_{N_{sat}} - \mathbf{P})\mathbf{H}\boldsymbol{\Delta}\mathbf{x} &= \mathbf{I}_{N_{sat}}\mathbf{H}\boldsymbol{\Delta}\mathbf{x} - \mathbf{H}(\mathbf{H}^T\mathbf{H})^{-1}\mathbf{H}^T\mathbf{H}\boldsymbol{\Delta}\mathbf{x} \\ &= \mathbf{I}_{N_{sat}}\mathbf{H}\boldsymbol{\Delta}\mathbf{x} - \mathbf{H}(\mathbf{H}^T\mathbf{H})^{-1}(\mathbf{H}^T\mathbf{H})\boldsymbol{\Delta}\mathbf{x} = \mathbf{I}_{N_{sat}}\mathbf{H}\boldsymbol{\Delta}\mathbf{x} - \mathbf{H}\mathbf{I}_4\boldsymbol{\Delta}\mathbf{x} = 0\end{aligned}$$

where $\mathbf{I}_4$ is the unit 4×4 matrix.

Therefore, the relationship (3.12), i.e.

$$\hat{\boldsymbol{\varepsilon}}_{rr} = (\mathbf{I}_{N_{sat}} - \mathbf{P})\boldsymbol{\varepsilon}$$

is proved. □

The statistical characteristics of $\hat{\boldsymbol{\varepsilon}}_{rr}$ are not easily derived from (3.12), since a correlation among the elements of the vector is present. An elegant way to overcome this problem is to introduce the parity method, which leads to a test statistic, which is equivalent to r_{LS}, but much easier from the point of view of the statistical characterization.

3.2.2 *Parity Method*

The parity method, [3, 6–8], does not represent a real alternative to the method described in Sect. 3.2.1, as it leads to the definition of a test statistic, which coincides with the one in (3.8). However, the method is interesting as it allows us to do some theoretical considerations and understand the normalization in (3.10).

The key point of the method is based on the idea to write the $(N_{sat} \times 4)$ matrix $\mathbf{H}$, the Jacobian of the navigation equations $\boldsymbol{\Delta}_\rho = \mathbf{H}\boldsymbol{\Delta}\mathbf{x} + \boldsymbol{\varepsilon}$ given in (2.7), as the product of two matrices. In the mathematical discipline of linear algebra, this factorization is called matrix decomposition, and it is used to implement efficient algorithms for solving a system of linear equations. The parity method uses the so-called QR decomposition, [9], since this decomposition leads to the definition of a test statistic, easily analysed from a theoretical point of view.

The main points of the method are the following.

- The QR decomposition is applied to the matrix $\mathbf{H}$ obtaining

$$\mathbf{H} = \mathbf{Q}\mathbf{R} \tag{3.14}$$

- The matrix $\mathbf{R}$ can be written in the form

$$\mathbf{R} = \begin{bmatrix} \mathbf{R}_1 \\ \mathbf{0} \end{bmatrix}$$

where $\mathbf{R}$ is a $N_{sat} \times 4$, and $\mathbf{R}_1$ is a 4×4 upper triangular matrix.
- The matrix $\mathbf{Q}$ is $N_{sat} \times N_{sat}$; its columns are orthonormal (i.e. $\mathbf{Q}^T\mathbf{Q} = \mathbf{I}$) and contains two contributions $\mathbf{Q}_1$ e $\mathbf{Q}_2$, in the form $\mathbf{Q} = [\mathbf{Q}_1\mathbf{Q}_2]$, where $\mathbf{Q}_1$ is $N_{sat} \times 4$, and $\mathbf{Q}_2$ is $N_{sat} \times (N_{sat} - 4)$, and $\mathbf{Q}_1^T\mathbf{Q}_1 = \mathbf{I}$, $\mathbf{Q}_2^T\mathbf{Q}_2 = \mathbf{I}$. The matrix $\mathbf{Q}_1$ can be used to evaluate the LS estimate of $\boldsymbol{\Delta}\mathbf{x}$, as $\hat{\boldsymbol{\Delta}}\mathbf{x} = \mathbf{R}_1^{-1}(\mathbf{Q}_1^T)\boldsymbol{\Delta}_\rho$, since it is possible to show that

$$\mathbf{R}_1^{-1}\mathbf{Q}_1^T = (\mathbf{H}^T\mathbf{H})^{-1}\mathbf{H}^T \tag{3.15}$$

Therefore, the expression $\hat{\boldsymbol{\Delta}}\mathbf{x} = \mathbf{R}_1^{-1}(\mathbf{Q}_1^T)\boldsymbol{\Delta}_\rho$ is equivalent to the LS solution given in (2.11) and can be used to compute the LS estimate of the position coordinates.

Proof of (3.15), The first step in the demonstration is to explicitly compute the matrix $\mathbf{H}^T\mathbf{H}$ as a function of $\mathbf{Q}$ and $\mathbf{R}$

$$\mathbf{H}^T\mathbf{H} = \mathbf{R}^T\mathbf{Q}^T\mathbf{Q}\mathbf{R} = \mathbf{R}^T\mathbf{R}$$

and since

$$\mathbf{R}^T\mathbf{R} = \left[\mathbf{R}_1^T|\mathbf{0}^T\right]\begin{bmatrix} \mathbf{R}_1 \\ \mathbf{0} \end{bmatrix} = \mathbf{R}_1^T\mathbf{R}_1$$

we can derive that

$$\mathbf{H}^T\mathbf{H} = \mathbf{R}_1^T\mathbf{R}_1$$

from which it obviously follows that

$$(\mathbf{H}^T\mathbf{H})^{-1} = (\mathbf{R}_1^T\mathbf{R}_1)^{-1} = (\mathbf{R}_1)^{-1}(\mathbf{R}_1^T)^{-1}$$

We can finally derive the equivalence between $(\mathbf{H}^T\mathbf{H})^{-1}\mathbf{H}^T$ and $\mathbf{R}_1^{-1}(\mathbf{Q}_1^T)$. In fact, since

$$\mathbf{H}^T = \mathbf{R}^T\mathbf{Q}^T = \left[\mathbf{R}_1^T|\mathbf{0}^T\right]\begin{bmatrix} \mathbf{Q}_1^T \\ \mathbf{Q}_2^T \end{bmatrix} = \left[\mathbf{R}_1^T\mathbf{Q}_1^T + \mathbf{0}^T\right]$$

we can write that

$$(\mathbf{H}^T\mathbf{H})^{-1}\mathbf{H}^T = (\mathbf{R}_1)^{-1}(\mathbf{R}_1^T)^{-1}\left[\mathbf{R}_1^T\mathbf{Q}_1^T\right] = \mathbf{R}_1^{-1}\mathbf{Q}_1^T \tag{3.16}$$

□

- The matrix $\mathbf{Q}_2$ can be used to introduce a vector of $N_{sat} - 4$ elements $\mathbf{p} = \mathbf{Q}_2^T \boldsymbol{\Delta}_\rho$, called *parity vector*, which takes into account the errors in the estimate of the position coordinates, since it depends on the term $\boldsymbol{\varepsilon}$ in the navigation equations. In fact, it is possible to show that

$$\mathbf{p} = \mathbf{Q}_2^T \boldsymbol{\varepsilon} \tag{3.17}$$

Proof of (3.17). Starting from the equations $\mathbf{p} = \mathbf{Q}_2^T \boldsymbol{\Delta}_\rho$, and $\boldsymbol{\Delta}_\rho = \mathbf{H}\boldsymbol{\Delta}\mathbf{x} + \boldsymbol{\varepsilon}$, and recalling (3.14), we can write

$$\mathbf{p} = \mathbf{Q}_2^T (\mathbf{H}\boldsymbol{\Delta}\mathbf{x} + \boldsymbol{\varepsilon}) = \mathbf{Q}_2^T \mathbf{Q}\mathbf{R}\boldsymbol{\Delta}\mathbf{x} + \mathbf{Q}_2^T \boldsymbol{\varepsilon}$$

We observe that

$$\mathbf{Q}_2^T \mathbf{Q} = \mathbf{Q}_2^T [\mathbf{Q}_1 \mathbf{Q}_2] = [\mathbf{0}\ \mathbf{I}_{N_{sat}-4}]$$

and

$$[\mathbf{0}\ \mathbf{I}_{N_{sat}-4}]\mathbf{R} = [\mathbf{0}\ \mathbf{I}_{N_{sat}-4}] \begin{bmatrix} \mathbf{R}_1 \\ \mathbf{0} \end{bmatrix} = \mathbf{0}$$

Therefore, the relationship $\mathbf{p} = \mathbf{Q}_2^T \boldsymbol{\varepsilon}$ is proved. □

At this point, it is possible to introduce a test statistic of the type:

$$r_{\mathrm{P}} = \sqrt{\frac{\mathbf{p}^T \mathbf{p}}{N_{sat} - 4}} = \sqrt{\frac{\varepsilon_P}{N_{sat} - 4}} \tag{3.18}$$

However, this is not a true alternative to the test statistic r_{LS} given in (3.10). In fact, r_{P}, and r_{LS} are equivalent, since

$$\mathbf{p}^T \mathbf{p} = \hat{\boldsymbol{\varepsilon}}_{rr}^T \hat{\boldsymbol{\varepsilon}}_{rr} \tag{3.19}$$

Proof of (3.19). Since the matrix $\mathbf{Q}$ is orthogonal, the following equalities hold

$$\mathbf{Q}\mathbf{Q}^T = \mathbf{I} = \mathbf{Q}^T \mathbf{Q}$$

In particular,

$$\mathbf{Q}\mathbf{Q}^T = [\mathbf{Q}_1 | \mathbf{Q}_2] \begin{bmatrix} \mathbf{Q}_1^T \\ \mathbf{Q}_2^T \end{bmatrix} = \mathbf{Q}_1 \mathbf{Q}_1^T + \mathbf{Q}_2 \mathbf{Q}_2^T = \mathbf{I}$$

Moreover, we can show that $\mathbf{H}(\mathbf{H}^T\mathbf{H}^{-1})\mathbf{H}^T = \mathbf{Q}_1\mathbf{Q}_1^T$. In fact, from (3.16),

$$\mathbf{H}(\mathbf{H}^T\mathbf{H}^{-1})\mathbf{H}^T = \mathbf{H}\mathbf{R}_1^{-1}\mathbf{Q}_1^T$$

and knowing that an equivalent expression for the matrix $\mathbf{H}$ is

$$\mathbf{H} = [\mathbf{Q}_1 | \mathbf{Q}_2] \begin{bmatrix} \mathbf{R}_1 \\ \mathbf{0} \end{bmatrix} = \mathbf{Q}_1 \mathbf{R}_1$$

we can derive

$$\mathbf{H}\mathbf{R}_1^{-1}\mathbf{Q}_1^T = \mathbf{Q}_1\mathbf{R}_1\mathbf{R}_1^{-1}\mathbf{Q}_1^T = \mathbf{Q}_1\mathbf{Q}_1^T$$

Finally, we can show that $\mathbf{Q}_2\mathbf{Q}_2^T$ is equal to $\mathbf{I} - \mathbf{H}(\mathbf{H}^T\mathbf{H}^{-1})\mathbf{H}^T$

$$\mathbf{I} - \mathbf{H}(\mathbf{H}^T\mathbf{H}^{-1})\mathbf{H}^T = \mathbf{I} - \mathbf{Q}_1\mathbf{Q}_1^T = \mathbf{Q}_1\mathbf{Q}_1^T + \mathbf{Q}_2\mathbf{Q}_2^T - \mathbf{Q}_1\mathbf{Q}_1^T = \mathbf{Q}_2\mathbf{Q}_2^T$$

Therefore, (3.12) can be written as

$$\hat{\boldsymbol{\varepsilon}}_{rr} = (\mathbf{I}_{N_{sat}} - \mathbf{P})\boldsymbol{\varepsilon} = \mathbf{Q}_2\mathbf{Q}_2^T\boldsymbol{\varepsilon} = \mathbf{Q}_2\mathbf{p}$$

and then,

$$\hat{\boldsymbol{\varepsilon}}_{rr}^T\hat{\boldsymbol{\varepsilon}}_{rr} = (\mathbf{Q}_2\mathbf{p})^T\mathbf{Q}_2\mathbf{p} = \mathbf{p}^T\mathbf{Q}_2^T\mathbf{Q}_2\mathbf{p} = \mathbf{p}^T\mathbf{p}$$

□

The advantage of this method is that the statistical characteristics of ε_P (and then of r_{LS}) are easily derived, since the columns of the $\mathbf{Q}$ matrix are orthonormal.

3.2.2.1 Statistical Characteristics of the Parity Vector

The PDF of $\mathbf{p} = \mathbf{Q}_2^T\boldsymbol{\varepsilon}$ is easily obtained when $\boldsymbol{\varepsilon}$ is a vector of independent zero-mean Gaussian random variables with variance $\sigma_{\varepsilon_i}^2, i = 1, 2, \ldots, N_{sat}$. In fact, in this case, $\mathbf{p}$ is a vector of Gaussian random variables, with mean

$$E\{\mathbf{p}\} = \mathbf{Q}_2^T E\{\boldsymbol{\varepsilon}\} = \mathbf{0}$$

and covariance matrix

$$E\{\mathbf{p}\mathbf{p}^T\} = E\{\mathbf{Q}_2^T\boldsymbol{\varepsilon}(\mathbf{Q}_2^T\boldsymbol{\varepsilon})^T\} = E\{\mathbf{Q}_2^T\boldsymbol{\varepsilon}\boldsymbol{\varepsilon}^T\mathbf{Q}_2\} = \mathbf{Q}_2^T E\{\boldsymbol{\varepsilon}\boldsymbol{\varepsilon}^T\}\mathbf{Q}_2$$

since in general $(\mathbf{AP})^T = \mathbf{P}^T\mathbf{A}^T$, [9].

If now all the variances $\sigma_{\varepsilon_i}^2$ are equal, $\boldsymbol{\varepsilon}$ can be written as $\boldsymbol{\varepsilon} = \sigma_\varepsilon\boldsymbol{\varepsilon}_0$, where $\boldsymbol{\varepsilon}_0$ contains independent zero-mean unit-variance Gaussian random variables, and

$$E\{\boldsymbol{\varepsilon}\boldsymbol{\varepsilon}^T\} = \sigma_\varepsilon^2 E\{\boldsymbol{\varepsilon}_0\boldsymbol{\varepsilon}_0^T\} = \sigma_\varepsilon^2\mathbf{I}$$

from which

$$E\{\mathbf{p}\mathbf{p}^T\} = \sigma_\varepsilon^2\mathbf{Q}_2^T\mathbf{I}\mathbf{Q}_2 = \sigma_\varepsilon^2\mathbf{I}$$

3.2.2.2 Statistical Characteristics of the Test Statistic

If all the variances $\sigma^2_{\varepsilon_i}$ are equal, the PDF of the squared value of the test statistic, r^2_P is immediately found. In fact, in this case, r^2_P is a random variable with a χ^2 distribution with $N_{sat} - 4$ degrees of freedom. It is known that this distribution describes a random variable of the type:

$$X = \sum_{i=1}^{k} X_i^2$$

where X_i are independent random variable with normal distribution $\mathcal{N}(0, 1)$, and $k = N_{sat} - 4$. The mathematical expression of the PDF is

$$f_X(x; k) = \begin{cases} \frac{1}{2^{k/2}\Gamma(k/2)} x^{k/2-1} e^{-x/2} & x \geq 0 \\ 0 & x < 0 \end{cases} \tag{3.20}$$

where

$$\Gamma(\alpha) = \int_0^{\infty} y^{\alpha-1} e^{-y} dy$$

If the variable X_i are independent random variable with normal distribution $\mathcal{N}(0, \sigma_x)$, we have a variable

$$X_\sigma = \sum_{i=1}^{k} \sigma_x^2 X_i^2 = \sigma_x^2 X$$

whose PDF, easily obtained from (3.20), is

$$f_X(x; k) = \begin{cases} \frac{1}{2^{k/2}\sigma_x^2\Gamma(k/2)} \left(\frac{x}{\sigma_x^2}\right)^{k/2-1} e^{-x/2\sigma_x^2} & x \geq 0 \\ 0 & x < 0 \end{cases} \tag{3.21}$$

In our case, the random variable ε_P has the structure of X_σ where σ_x^2 is the variance σ_ε^2 of the errors in the pseudorange measurements.

3.2.3 *Computation of the Decision Threshold for the AWGN Model*

To detect the presence of a faulty PVT solution, a fault detector is implemented, which compares the test statistic against a threshold, in order to verify which of the two possible hypotheses

$$\mathcal{H}_0 : \text{No fault is present (Nominal condition)}$$
$$\mathcal{H}_1 : \text{A fault is present (Fault condition)}$$

are true.

The key point now is the evaluation of the threshold. A method generally adopted to fix the threshold is to use the Neyman–Pearson approach to signal detection, [10], based on the choice of a target false alarm probability P_{FA}. A false alarm occurs when a non-nominal state is declared while the true state of the system is nominal. For further reading check "Kay, Fundamental of Statistical Signal Processing". This allows the analytical evaluation of the threshold, since in nominal conditions (AWGN model) the probability density function of ε_P is the chi-square distribution given in (3.21).

To find the threshold $T_{th,P}$ of r_P, the first step is to set a target false alarm probability $P_{FA,target}$. Then, the threshold is found from

$$P_{FA,target} = \Pr(r_P > T_{th,P})|\mathcal{H}_0)$$

The event $r_P > T_{th,P}$ corresponds to the event

$$\frac{\varepsilon_P}{N_{sat} - 4} > T_{th,P}^2$$

or equivalently

$$\varepsilon_{P,norm} > \frac{T_{th,P}^2 (N_{sat} - 4)}{\sigma_\varepsilon^2} = T_{th,\varepsilon}$$

where $\varepsilon_{P,norm} = \varepsilon_P / \sigma_\varepsilon^2$. At this point, the target false alarm probability can be written as

$$P_{FA,target} = \Pr(r_P > T_{th,P}|\mathcal{H}_0) = \Pr(\varepsilon_{P,norm} > T_{th,\varepsilon}|\mathcal{H}_0) = \int_{T_{th,\varepsilon}}^{\infty} f_{\varepsilon_{P,norm}}(\varepsilon) d\varepsilon$$

The value of $T_{th,\varepsilon}$ can be obtained by inverting this integral, from which the threshold $T_{th,P}$ can be found as

$$T_{th,P} = \sqrt{\frac{T_{th,\varepsilon}\sigma_\varepsilon^2}{N_{sat} - 4}} = \sigma_\varepsilon \sqrt{\frac{T_{th,\varepsilon}}{N_{sat} - 4}}$$

The diagram of $\sqrt{T_{th,\varepsilon}/(N_{sat} - 4)}$ versus the target false alarm probability is shown in Fig. 3.1. This diagram can be used to evaluate the threshold, given a target false alarm probability. This threshold is then used to claim that a fault is present when

$$r_{LS} = r_P > \sigma_\varepsilon \sqrt{\frac{T_{th,\varepsilon}}{N_{sat} - 4}}$$

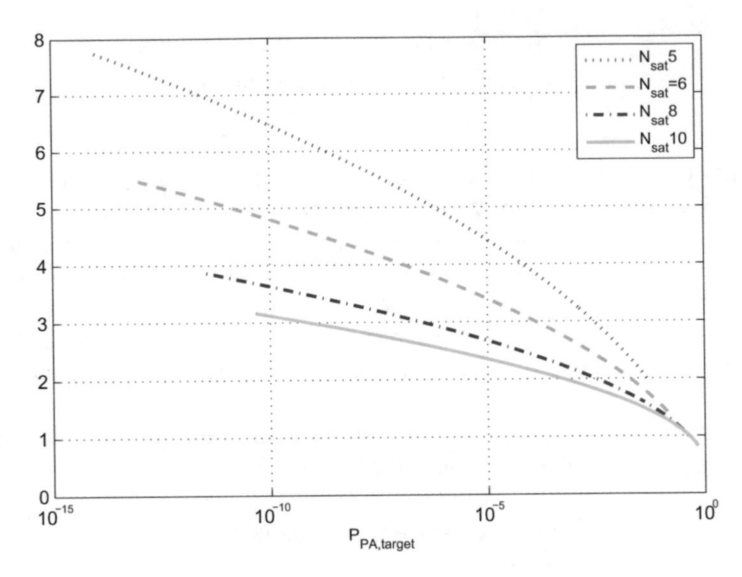

Fig. 3.1 Diagram of $\sqrt{T_{th,\varepsilon}/(N_{sat}-4)}$ versus the target false alarm probability

3.3 Fault Detection and Exclusion in the Position Domain

The FD methods seen in Sect. 3.2 (range residual and parity) work in the range domain and allows us to identify if faults are present in the computed position. The methods do not require any assumption of the number and type of faults and then are quite general and can be applied for any GNSS application. We consider now the problem of isolating the satellites with failures. This is a much more challenging task, which cannot be performed without making some assumptions on the number and type of faults.

In this section, we describe the method known as *solution separation test*, based on the hypothesis $\mathcal{H}_a$ that only a single satellite is faulty, and the fault is a bias in the pseudorange. In this case, the hypothesis $\mathcal{H}_a$ can be partitioned in N_{sat} mutually exclusive hypotheses $\mathcal{H}_{a,1}, \mathcal{H}_{a,2}, \ldots, \mathcal{H}_{a,n}$ that corresponds to the various single satellite fault cases (to simplify the notations, the symbol $\mathcal{H}_{a,n}$ is used instead of $\mathcal{H}_{a,N_{sat}}$). The overall set of considered hypotheses is thus

$$
\begin{aligned}
\mathcal{H}_0 &: \text{No fault is present (Nominal condition)}\\
\mathcal{H}_{a,1} &: \text{The } 1_{st} \text{ satellite (only), has a bias in the measured pseudorange}\\
\mathcal{H}_{a,2} &: \text{The } 2_{nd} \text{ satellite (only), has a bias in the measured pseudorange}\\
&\ldots\\
\mathcal{H}_{a,i} &: \text{The } i_{th} \text{ satellite (only), has a bias in the measured pseudorange}\\
&\ldots\\
\mathcal{H}_{a,n} &: \text{The } n_{th} \text{ satellite (only), has a bias in the measured pseudorange}
\end{aligned}
$$

For example, under hypothesis $\mathscr{H}_{a,n}$ (the bias is in the last measured pseudorange), the measurement error vector $\boldsymbol{\varepsilon}|_{\mathscr{H}_{a,n}}$ has the following distribution

$$\boldsymbol{\varepsilon}|_{\mathscr{H}_{a,n}} \sim N(\left[0\ 0\ 0 \cdots 0\ b\right]^T, \sigma_\varepsilon^2 \mathbf{I})$$

where b is the value of the bias and σ_ε^2 is the variance of the noise components.

Once a faulty satellite is identified, this can be excluded in the PVT computation. For this reason, any algorithm able to isolate a satellite fault is generally called method of FDE. However, also, different countermeasures can be adopted to take into account the presence of faults.

In this section, we will describe in details the method used to identify a single fault and treat the multi fault case as a generalization.

3.3.1 Solution Separation with a Single Fault

Consider, just for simplicity in explaining the solution separation method, that the only possible alternative to hypothesis $\mathscr{H}_0$ is hypothesis $\mathscr{H}_{a,n}$. The set of considered hypotheses is then

$$\begin{aligned} \mathscr{H}_0 &: \text{No fault is present (Nominal condition)} \\ \mathscr{H}_{a,n} &: \text{The } n_{th} \text{ satellite (only), has a bias in the measured pseudorange} \end{aligned}$$

In this case, the solution separation test becomes a method which decides if the estimated user position is under the hypothesis $\mathscr{H}_0$ or under the hypothesis $\mathscr{H}_{a,n}$. However, since the analysis of the statistical properties of the method is difficult, we will analyse the problem from a different perspective. We first build a MMSE estimator $\hat{b}$ for the bias b (under hypothesis $\mathscr{H}_{a,n}$) and compare the estimated value against a threshold for deciding between the hypothesis $\mathscr{H}_0$ and hypothesis $\mathscr{H}_{a,n}$. Then, we will show that the difference $\mathbf{d}$ between the estimated user position under hypothesis $\mathscr{H}_0$ (i.e. considering all the satellites in view) and under hypothesis $\mathscr{H}_{a,n}$ (i.e. considering all the satellites in view, except the one in the position N_{sat}) is equal to the estimator $\hat{b}$ multiplied by a known, geometry only dependent, vector. Demonstrating that $\mathbf{d}$ is an equivalent estimator for the bias gives the theoretical justification for the usage of this kind of test as a fault detection mechanism.

3.3.2 Bias Estimator

Under hypothesis $\mathscr{H}_{a,n}$, we can build the MMSE estimators of the unknown position vector $\boldsymbol{\Delta}\mathbf{x}$ and the unknown bias b. We can start our analysis by considering that the system of equations describing the measurement process under hypothesis $\mathscr{H}_{a,n}$ is

$$\boldsymbol{\Delta}_\rho = \mathbf{H}\boldsymbol{\Delta}\mathbf{x} + \boldsymbol{\varepsilon}|_{\mathscr{H}_{a,n}} \tag{3.22}$$

or, equivalently

$$\boldsymbol{\Delta}_\rho = \mathbf{H}\boldsymbol{\Delta}\mathbf{x} + \boldsymbol{\varepsilon}|_{\mathscr{H}_0} + \mathbf{b} \tag{3.23}$$

where

$$\boldsymbol{\varepsilon}|_{\mathscr{H}_0} \sim N(\mathbf{0}^T, \sigma_\varepsilon^2 \mathbf{I})$$

Equation (3.23) can be easily rewritten as

$$\boldsymbol{\Delta}_\rho = \left[\mathbf{H} \mid \mathbf{u}\right] \left[\frac{\boldsymbol{\Delta}\mathbf{x}}{b}\right] + \boldsymbol{\varepsilon} \tag{3.24}$$

where $\mathbf{u} = \left[0\ 0\ 0 \cdots 1\right]^T$. Defining the matrix $\mathbf{M} = \left[\mathbf{H} \mid \mathbf{u}\right]$, it is evident that we can construct the estimator for the position and the unknown bias in the usual way for a linear system

$$\left[\frac{\hat{\boldsymbol{\Delta}\mathbf{x}}|_{\mathscr{H}_{a,n}}}{\hat{b}}\right] = (\mathbf{M}^T\mathbf{M})^{-1}\mathbf{M}^T\boldsymbol{\Delta}_\rho \tag{3.25}$$

3.3.2.1 Separation of the Two Estimates in (3.25)

To separate the two estimates in (3.25), it is useful to conceptually partition the geometry matrix $\mathbf{H}$ in a submatrix $\mathbf{H}_M$ of dimension $(N_{sat} - 1) \times 4$ and its last row $\mathbf{h}$

$$\mathbf{H} = \left[\frac{\mathbf{H}_M}{\mathbf{h}}\right]$$

and to define a $N_{sat} \times 4$ matrix $\mathbf{H}_R$ as

$$\mathbf{H}_R = \left[\frac{\mathbf{H}_M}{\mathbf{0}}\right]$$

These two matrices $\mathbf{H}_M$ and $\mathbf{H}_R$ are now used to write (3.25) in another form. We start by noticing that

$$\mathbf{M}^T\mathbf{M} = \left[\frac{\mathbf{H}^T}{\mathbf{u}^T}\right]\left[\mathbf{H} \mid \mathbf{u}\right] = \left[\begin{array}{c|c} \mathbf{H}^T\mathbf{H} & \mathbf{h}^T \\ \hline \mathbf{h} & \mathbf{u}^T\mathbf{u} \end{array}\right] = \left[\begin{array}{c|c} \mathbf{H}^T\mathbf{H} & \mathbf{h}^T \\ \hline \mathbf{h} & 1 \end{array}\right]$$

Knowing the properties for the inverse of a block matrix [11], we can derive

$$(\mathbf{M}^T\mathbf{M})^{-1} = \left[\begin{array}{c|c} (\mathbf{H}^T\mathbf{H} - \mathbf{h}^T\mathbf{h})^{-1} & -(\mathbf{H}^T\mathbf{H} - \mathbf{h}^T\mathbf{h})^{-1}\mathbf{h}^T \\ \hline -\mathbf{h}(\mathbf{H}^T\mathbf{H} - \mathbf{h}^T\mathbf{h})^{-1} & 1 + \mathbf{h}(\mathbf{H}^T\mathbf{H} - \mathbf{h}^T\mathbf{h})^{-1}\mathbf{h}^T \end{array}\right]$$

Noticing that

$$(\mathbf{H}^T\mathbf{H} - \mathbf{h}^T\mathbf{h})^{-1} = ({\mathbf{H}_R}^T\mathbf{H}_R)^{-1}$$

it is then possible to finally derive that

$$(\mathbf{M}^T\mathbf{M})^{-1}\mathbf{M}^T = \left[\begin{array}{c} ({\mathbf{H}_R}^T\mathbf{H}_R)^{-1}{\mathbf{H}_R}^T \\ \hline \mathbf{u}^T - \mathbf{h}({\mathbf{H}_R}^T\mathbf{H}_R)^{-1}{\mathbf{H}_R}^T \end{array}\right]$$

The estimated position $\hat{\boldsymbol{\Delta}}\mathbf{x}|_{\mathscr{H}_{a,n}}$ under the hypothesis $H_{a,n}$ is thus simply calculated by excluding the last pseudorange from the set of measurements, that is

$$\hat{\boldsymbol{\Delta}}\mathbf{x}|_{\mathscr{H}_{a,n}} = ({\mathbf{H}_R}^T\mathbf{H}_R)^{-1}{\mathbf{H}_R}^T\boldsymbol{\Delta}_\rho = ({\mathbf{H}_M}^T\mathbf{H}_M)^{-1}{\mathbf{H}_M}^T\boldsymbol{\Delta}_{\rho M} \tag{3.26}$$

where $\boldsymbol{\Delta}_{\rho M}$ is a column vector $(N_{sat} - 1) \times 1$ that contains the values of $\boldsymbol{\Delta}_\rho$ from row 1 up to row $N_{sat} - 1$.

Interestingly, the estimated bias $\hat{b}$ can be expressed as the difference between the last measured pseudorange and the scalar product between the estimated position under fault condition and the vector $\mathbf{h}$:

$$\hat{b} = (\mathbf{u}^T - \mathbf{h}({\mathbf{H}_R}^T\mathbf{H}_R)^{-1}{\mathbf{H}_R}^T)\boldsymbol{\Delta}_\rho = \Delta_{\rho,N_{sat}} - \mathbf{h}\hat{\boldsymbol{\Delta}}\mathbf{x}|_{\mathscr{H}_{a,n}} \tag{3.27}$$

This equation shows how to build an estimator $\hat{b}$ for the bias under the fault hypothesis $\mathscr{H}_{a,n}$. This value can thus be compared against a threshold for deciding if the last satellite is faulty or healthy. In fact, it can be easily shown that the estimated value $\hat{b}$ is equal to the true bias b plus a Gaussian noise component

$$\hat{b} = b + n \quad n \sim N(0, \sigma_\varepsilon^2(1 + ||\mathbf{h}({\mathbf{H}_M}^T\mathbf{H}_M)^{-1}{\mathbf{H}_M}^T||^2)) \tag{3.28}$$

This estimated value can now be compared against a threshold, obtained by fixing a false alarm probability.

3.3.2.2 Comparison with the Standard Method

In the GNSS literature, a different estimator is generally used for the SS-based detection, based on the difference between the estimated position under hypothesis $\mathscr{H}_0$ and the estimated position under hypothesis $\mathscr{H}_{a,n}$, i.e. $\hat{\boldsymbol{\Delta}}\mathbf{x}|_{\mathscr{H}_0} - \hat{\boldsymbol{\Delta}}\mathbf{x}|_{\mathscr{H}_{a,n}}$. The key point for understanding why this difference vector can be used for hypothesis decision is that the nominal condition estimated position can be rewritten as the sum of the alternative hypothesis condition estimated position and a function of the estimated bias. In particular, it is possible to show that

$$\hat{\boldsymbol{\Delta}}\mathbf{x}|_{\mathscr{H}_0} - \hat{\boldsymbol{\Delta}}\mathbf{x}|_{\mathscr{H}_{a,n}} = (\mathbf{H}^T\mathbf{H})^{-1}\mathbf{h}^T\hat{b} \tag{3.29}$$

Therefore, the difference between the two estimated user positions is a vector that can be used for the estimation of the bias and consequently for the detection of a fault.

Proof of (3.29). The nominal condition estimated position is calculated as

$$\hat{\boldsymbol{\Delta}}\mathbf{x}|_{\mathscr{H}_0} = (\mathbf{H}^T\mathbf{H})^{-1}\mathbf{H}^T\boldsymbol{\Delta}_\rho$$

where the term $(\mathbf{H}^T\mathbf{H})^{-1}$ can be rewritten as

$$\begin{aligned}
&(\mathbf{H}^T\mathbf{H})^{-1} = (\mathbf{H}^T\mathbf{H})^{-1}(\mathbf{H}_R^T\mathbf{H}_R)(\mathbf{H}_R^T\mathbf{H}_R)^{-1} = \\
&(\mathbf{H}^T\mathbf{H})^{-1}(\mathbf{H}^T\mathbf{H} - \mathbf{h}^T\mathbf{h})(\mathbf{H}_R^T\mathbf{H}_R)^{-1} = \\
&(\mathbf{I} - (\mathbf{H}^T\mathbf{H})^{-1}\mathbf{h}^T\mathbf{h})(\mathbf{H}_R^T\mathbf{H}_R)^{-1}
\end{aligned}$$

and the term $(\mathbf{H}^T\mathbf{H})^{-1}\mathbf{H}^T$ as

$$\begin{aligned}
&(\mathbf{H}^T\mathbf{H})^{-1}\mathbf{H}^T = (\mathbf{I} - (\mathbf{H}^T\mathbf{H})^{-1}\mathbf{h}^T\mathbf{h})(\mathbf{H}_R^T\mathbf{H}_R)^{-1}\left[\mathbf{H}_M^T \mid \mathbf{h}^T\right] = \\
&\left[(\mathbf{I} - (\mathbf{H}^T\mathbf{H})^{-1}\mathbf{h}^T\mathbf{h})(\mathbf{H}_R^T\mathbf{H}_R)^{-1}\mathbf{H}_M^T \mid (\mathbf{H}^T\mathbf{H})^{-1}\mathbf{h}^T\right]
\end{aligned}$$

Finally, an equivalent expression for $\hat{\boldsymbol{\Delta}}\mathbf{x}|_{\mathscr{H}_0}$ is

$$\begin{aligned}
&\hat{\boldsymbol{\Delta}}\mathbf{x}|_{\mathscr{H}_0} = (\mathbf{H}^T\mathbf{H})^{-1}\mathbf{H}^T\boldsymbol{\Delta}_\rho = \\
&\left[(\mathbf{I} - (\mathbf{H}^T\mathbf{H})^{-1}\mathbf{h}^T\mathbf{h})(\mathbf{H}_R^T\mathbf{H}_R)^{-1}\mathbf{H}_M^T \mid (\mathbf{H}^T\mathbf{H})^{-1}\mathbf{h}^T\right]\boldsymbol{\Delta}_\rho = \\
&\left[(\mathbf{I} - (\mathbf{H}^T\mathbf{H})^{-1}\mathbf{h}^T\mathbf{h})(\mathbf{H}_R^T\mathbf{H}_R)^{-1}\mathbf{H}_M^T \mid (\mathbf{H}^T\mathbf{H})^{-1}\mathbf{h}^T\right]\left[\frac{\boldsymbol{\Delta}_{\rho M}}{\Delta_{\rho,N_{sat}}}\right] = \\
&(\mathbf{I} - (\mathbf{H}^T\mathbf{H})^{-1}\mathbf{h}^T\mathbf{h})\hat{\boldsymbol{\Delta}}\mathbf{x}|_{\mathscr{H}_{a,n}} + (\mathbf{H}^T\mathbf{H})^{-1}\mathbf{h}^T\Delta_{\rho,N_{sat}} = \\
&\hat{\boldsymbol{\Delta}}\mathbf{x}|_{\mathscr{H}_{a,n}} - (\mathbf{H}^T\mathbf{H})^{-1}\mathbf{h}^T(\mathbf{H}\hat{\boldsymbol{\Delta}}\mathbf{x}|_{\mathscr{H}_{a,n}} - \Delta_{\rho,N_{sat}}) = \\
&\hat{\boldsymbol{\Delta}}\mathbf{x}|_{\mathscr{H}_{a,n}} + (\mathbf{H}^T\mathbf{H})^{-1}\mathbf{h}^T\hat{b}
\end{aligned}$$

from which

$$\hat{\boldsymbol{\Delta}}\mathbf{x}|_{\mathscr{H}_0} - \hat{\boldsymbol{\Delta}}\mathbf{x}|_{\mathscr{H}_{a,n}} = (\mathbf{H}^T\mathbf{H})^{-1}\mathbf{h}^T\hat{b}$$

□

3.3.2.3 Generalization of the Method in Case of Single Fault

In a real application, obviously, the set of considered hypotheses is $\mathscr{H}_{a,1}, \mathscr{H}_{a,2}, \ldots, \mathscr{H}_{a,n}$, i.e. the bias can be in any of the pseudoranges. Usually, N_{sat} different tests are performed, one for each possible fault hypothesis, and according to the test results, the faulty satellite is excluded.

3.3.3 Solution Separation Test with a Generic Number of Faults

The generic solution separation (SS) test associated to a generic fault hypothesis is based on the difference between the so-called all-in-view position estimation, $\hat{\boldsymbol{\Delta}\mathbf{x}}^{(T)}$, and the estimated position $\hat{\boldsymbol{\Delta}\mathbf{x}}^{(S)}$ obtained by using a reduced subset (the one associated to the considered fault hypothesis), that is

$$\mathbf{d}^{(0\rightarrow F)} = \hat{\boldsymbol{\Delta}\mathbf{x}}^{(T)} - \hat{\boldsymbol{\Delta}\mathbf{x}}^{(S)} \tag{3.30}$$

where T is the complete set of satellites and S is the set difference between the all satellite set T and the set of faulty satellites F corresponding to the considered fault mode hypothesis; then,

$$S = T \setminus F \tag{3.31}$$

This notation can be generalized as

$$\mathbf{d}^{(A\rightarrow B)} = \hat{\boldsymbol{\Delta}\mathbf{x}}^{(T\setminus A)} - \hat{\boldsymbol{\Delta}\mathbf{x}}^{(T\setminus B)} \tag{3.32}$$

where A and B are the set of excluded satellites.

The position solution with the reduced subset S is computed as

$$\hat{\boldsymbol{\Delta}\mathbf{x}}^{(S)} = \mathbf{G}_S \mathbf{H}_S^T \boldsymbol{\Delta\rho} \tag{3.33}$$

where S is a generic subset of at least four elements of the set T of all the measured pseudoranges, $\mathbf{H}_S$ is the corresponding reduced geometry matrix, and $\mathbf{G}_S = (\mathbf{H}_S^T \mathbf{H}_S)^{-1}$. In formal terms,

$$\mathbf{H}_S = \mathbf{R}_S \mathbf{H} \tag{3.34}$$

where

$$\mathbf{R}_S = \mathrm{diag}(\boldsymbol{v}_S) \tag{3.35}$$

and

$$v_{S,q} = \begin{cases} 1 \text{ if } T_q \in S \\ 0 \text{ otherwise} \end{cases} \tag{3.36}$$

For each possible fault mode ($i = 1, \ldots, N_{fault}$), we have a well defined fault tolerant subset S_i, a matrix $\mathbf{H}_{S_i}$, and an associated subset fault tolerant solution $\hat{\boldsymbol{\Delta}}\mathbf{x}^{(S_i)}$, where the subset S_i is the set difference between the all-in-view subset T and the set of faulty satellites F_i corresponding to i_{th} fault mode, that is

$$S_i = T \setminus F_i \tag{3.37}$$

For every fault mode and for the three spatial estimates (x, y, z), the classical SS algorithm, described in detail in [12], uses the following test:

$$\frac{|\hat{\Delta x}_k^{(T)} - \hat{\Delta x}_k^{(S_i)}|}{T_k^{(0 \to T \setminus S_i)}} \lessgtr 1 \quad k = 1, 2, 3 \tag{3.38}$$

where $T_k^{(0 \to T \setminus S_i)}$ is

$$T_k^{(0 \to T \setminus S_i)} = \sqrt{(\mathbf{L}^{(0 \to T \setminus S_i)} \mathbf{L}^{(0 \to T \setminus S_i)^T})_{k,k}} \tag{3.39}$$

and

$$\mathbf{L}^{(0 \to T \setminus S_i)} = ((\mathbf{H}^T\mathbf{H})^{-1}\mathbf{H}^T - (\mathbf{H}_{S_i}^T\mathbf{H}_{S_i})^{-1}\mathbf{H}_{S_i}^T) \tag{3.40}$$

Notice that the k-sigma factors introduced in both [12, 13] are neglected in (3.39). In fact, in these documents, the complete definition of the threshold is

$$T_k^{(0 \to T \setminus S_i)} = \sqrt{(\mathbf{L}^{(0 \to T \setminus S_i)} \mathbf{L}^{(0 \to T \setminus S_i)^T})_{k,k}} K_{fa,k} \sigma_\epsilon \tag{3.41}$$

where the quantities $K_{fa,k}$, $k = 1, 2, 3$ are easily derived as explained in [12] and σ_ϵ is the standard deviation of the measurement error (assumed to be equal among all the pseudoranges). Since their presence does not influence the functioning of our proposed algorithm and since the notation is already heavy enough, we decided to neglect them. In Sect. 3.5, a low-complexity implementation of the test is presented.

3.4 Comparison Between Methods in the Range and in the Position Domains

In this section, we analyse the relationships between the range residual vector $\hat{\boldsymbol{\varepsilon}}_{rr}$ and the difference of estimated positions $\hat{\boldsymbol{\Delta}}\mathbf{x}|_{\mathscr{H}_0} - \hat{\boldsymbol{\Delta}}\mathbf{x}|_{\mathscr{H}_{a,n}}$. Again, in the derivation, we will assume that the set of considered hypotheses is

$\mathscr{H}_0$: No fault is present (Nominal condition)

$\mathscr{H}_{a,n}$: The n_{th} satellite (that is the satellite in position N_{sat}) has a bias in the measured pseudorange

We have seen in (3.29) that the delta position estimate is proportional to the estimated bias trough a known row vector $(\mathbf{H}^T\mathbf{H})^{-1}\mathbf{h}^T$ that is only geometry dependent, where the estimated bias, given in (3.27), is $\hat{b} = \Delta_{\rho,N_{sat}} - \mathbf{h}\hat{\boldsymbol{\Delta}}\mathbf{x}|_{\mathscr{H}_{a,n}}$.

Starting from (3.27), and substituting the expression of $\boldsymbol{\Delta}_\rho$ given in (3.23), it is possible to show that the relationship between the estimated bias, the error vector $\boldsymbol{\varepsilon} = \boldsymbol{\varepsilon}|_{\mathscr{H}_0}$, and the true bias vector is

$$\hat{b} = \mathbf{u}^T(\mathbf{I} - \mathbf{H}(\mathbf{H}_R^T\mathbf{H}_R)^{-1}\mathbf{H}_R^T)(\boldsymbol{\varepsilon} + \mathbf{b}) \tag{3.42}$$

Concerning the range residual test, we observe that the last element of the range residual vector $\hat{\boldsymbol{\varepsilon}}_{rr}$ can be written in the form

$$\hat{\varepsilon}_{rr,N_{sat}} = \mathbf{u}^T\hat{\boldsymbol{\varepsilon}}_{rr} = \mathbf{u}^T(\mathbf{I} - \mathbf{H}(\mathbf{H}^T\mathbf{H})^{-1}\mathbf{H}^T)(\boldsymbol{\varepsilon} + \mathbf{b}) \tag{3.43}$$

obtained from (3.12), including a bias in the error vector.

From (3.42) and (3.43), it is possible to derive that there is a linear relationship between $\hat{\varepsilon}_{rr,N_{sat}}$ and $\hat{b}$, only geometry dependent, that is

$$\hat{\varepsilon}_{rr,N_{sat}} = \frac{1}{1 + \mathbf{h}(\mathbf{H}_R^T\mathbf{H}_R)^{-1}\mathbf{h}^T}\hat{b} \tag{3.44}$$

Proof of (3.44), We start by substituting (3.42) in (3.44):

$$\hat{\varepsilon}_{rr,N_{sat}} = \frac{\mathbf{u}^T(\mathbf{I} - \mathbf{H}(\mathbf{H}_R^T\mathbf{H}_R)^{-1}\mathbf{H}_R^T)}{1 + \mathbf{h}(\mathbf{H}_R^T\mathbf{H}_R)^{-1}\mathbf{h}^T}(\boldsymbol{\varepsilon} + \mathbf{b})$$

where $\mathbf{h}(\mathbf{H}_R^T\mathbf{H}_R)^{-1}\mathbf{h}^T$ is a scalar which can be written as

$$\mathbf{h}(\mathbf{H}_R^T\mathbf{H}_R)^{-1}\mathbf{h}^T = k$$

Therefore, to prove (3.44), we have to prove that

$$(1 + k)\mathbf{u}^T(\mathbf{I} - \mathbf{H}(\mathbf{H}^T\mathbf{H})^{-1}\mathbf{H}^T) = \mathbf{u}^T(\mathbf{I} - \mathbf{H}(\mathbf{H}_R^T\mathbf{H}_R)^{-1}\mathbf{H}_R^T)$$

The lefthand side of this equation, written in the form

$$(1+k)\mathbf{u}^T(\mathbf{I}-\mathbf{H}(\mathbf{H}^T\mathbf{H})^{-1}\mathbf{H}^T) = (1+k)(\mathbf{u}^T - \mathbf{h}(\mathbf{H}^T\mathbf{H})^{-1}\mathbf{H}^T)$$

can be rewritten, using the expression of $(\mathbf{H}^T\mathbf{H})^{-1}$ given by the Sherma–Morrison formula [11]

$$(\mathbf{H}^T\mathbf{H})^{-1} = \left(\mathbf{I} - \frac{(\mathbf{H}_R^T\mathbf{H}_R)^{-1}\mathbf{h}^T\mathbf{h}}{1+\mathbf{h}(\mathbf{H}_R^T\mathbf{H}_R)^{-1}\mathbf{h}^T}\right)(\mathbf{H}_R^T\mathbf{H}_R)^{-1} = \left(\mathbf{I} - \frac{(\mathbf{H}_R^T\mathbf{H}_R)^{-1}\mathbf{h}^T\mathbf{h}}{1+k}\right)(\mathbf{H}_R^T\mathbf{H}_R)^{-1}$$

which allows us to write the quantity $(\mathbf{H}^T\mathbf{H})^{-1}$ as a function of $\mathbf{H}_R, \mathbf{h}$. We obtain

$$\begin{aligned}
&(1+k)(\mathbf{u}^T - \mathbf{h}(\mathbf{H}^T\mathbf{H})^{-1}\mathbf{H}^T) = (1+k)(\mathbf{u}^T - \mathbf{h}\left(\mathbf{I} - \frac{(\mathbf{H}_R^T\mathbf{H}_R)^{-1}\mathbf{h}^T\mathbf{h}}{1+k}\right)(\mathbf{H}_R^T\mathbf{H}_R)^{-1}\mathbf{H}^T) = \\
&(1+k)\mathbf{u}^T - (1+k)\mathbf{h}\left(\mathbf{I} - \frac{(\mathbf{H}_R^T\mathbf{H}_R)^{-1}\mathbf{h}^T\mathbf{h}}{1+k}\right)(\mathbf{H}_R^T\mathbf{H}_R)^{-1}\mathbf{H}^T) = \\
&(1+k)\mathbf{u}^T - (1+k)\mathbf{h}(\mathbf{H}_R^T\mathbf{H}_R)^{-1}\mathbf{H}^T + (1+k)\mathbf{h}\frac{(\mathbf{H}_R^T\mathbf{H}_R)^{-1}\mathbf{h}^T\mathbf{h}}{1+k}(\mathbf{H}_R^T\mathbf{H}_R)^{-1}\mathbf{H}^T \\
&(1+k)\mathbf{u}^T - (1+k)\mathbf{h}(\mathbf{H}_R^T\mathbf{H}_R)^{-1}\mathbf{H}^T + k\mathbf{h}(\mathbf{H}_R^T\mathbf{H}_R)^{-1}\mathbf{H}^T = \\
&(1+k)\mathbf{u}^T - \mathbf{h}(\mathbf{H}_R^T\mathbf{H}_R)^{-1}\mathbf{H}^T = \\
&(1+k)\mathbf{u}^T - \mathbf{h}(\mathbf{H}_R^T\mathbf{H}_R)^{-1}(\mathbf{H}_R^T + [\mathbf{0}|\mathbf{h}^T]) = \\
&(1+k)\mathbf{u}^T - \mathbf{h}(\mathbf{H}_R^T\mathbf{H}_R)^{-1}\mathbf{H}_R^T - [\mathbf{0}|\mathbf{h}(\mathbf{H}_R^T\mathbf{H}_R)^{-1}\mathbf{h}^T] = \\
&(1+k)\mathbf{u}^T - \mathbf{h}(\mathbf{H}_R^T\mathbf{H}_R)^{-1}\mathbf{H}_R^T - [\mathbf{0}|k] = \\
&\mathbf{u}^T - \mathbf{h}(\mathbf{H}_R^T\mathbf{H}_R)^{-1}\mathbf{H}_R^T
\end{aligned}$$

□

The conclusion is that also $\hat{\varepsilon}_{rr,N_{sat}}$ is an equivalent estimator of b.

The natural question that can arises in reader's mind is the following: Why do we have with the residual chi-square test to consider all the vector components of the residual vector $\hat{\boldsymbol{\varepsilon}}_{rr}$ while all the information is contained in its last element? The answer is hidden in the assumption we made about the nominal and alternative hypothesis $\mathscr{H}_0$ and $\mathscr{H}_{a,n}$. In fact, the only alternative hypothesis we made was $\mathscr{H}_{a,n}$ that stated that the only pseudorange with a bias was the last one.

If we extend the set of hypotheses from $\mathscr{H}_0, \mathscr{H}_{a,n}$ to $\mathscr{H}_0, \mathscr{H}_{a_1}, \mathscr{H}_{a_2}, \mathscr{H}_{a_3}, ..\mathscr{H}_{a_n}$, where the generic alternative hypothesis $\mathscr{H}_{a_i}$ refers to the ith pseudorange having a bias, and we build a set of MMSE estimators $\hat{b}^{(i)}$ as done in the previous section, then it is possible to show that:

- the difference between the nominal condition estimated position and the estimated position under hypothesis H_{a_i}, the vector $\hat{\boldsymbol{\Delta}}\mathbf{x}|_{\mathscr{H}_0} - \hat{\boldsymbol{\Delta}}\mathbf{x}|_{\mathscr{H}_{a,i}}$, is proportional to the estimated quantity $\hat{b}^{(i)}$ through a geometry only dependent factor
- the norm of the residual vector $\hat{\boldsymbol{\varepsilon}}_{rr}$ is a linear combination of the square of the estimated bias under the different hypothesis

$$||\hat{\boldsymbol{\varepsilon}}_{rr}||^2 = \sum_i c_i(\hat{b}^{(i)})^2$$

It is evident from these two last properties that in a real-life positioning applications, where the set of considered hypotheses is generally $\mathscr{H}_0, \mathscr{H}_{a,1}, \mathscr{H}_{a,2}, \mathscr{H}_{a,3}, \ldots \mathscr{H}_{a,n}$, the chi-square test can be used to determine whether a generic pseudorange contains a bias, while the solution separation tests can be used to decide if the receiver is working in nominal conditions or not, and if not, what is the most likely biased pseudorange.

3.5 Residual and Solution Separation Tests: Geometric Interpretation and Efficient Implementation

The purpose of this section is to give a general picture of the fault detection problem and exclusion from a geometrical point of view and to introduce an efficient implementation of the solution separation test. We first consider the range residual test method and then the solution separation test.

To simplify the notation, we write $\boldsymbol{\Delta}_\rho = \mathbf{H}\boldsymbol{\Delta}\mathbf{x} + \boldsymbol{\varepsilon}|_{\mathscr{H}_0} + \mathbf{b}$ as

$$\boldsymbol{\Delta}_\rho = \mathbf{H}\boldsymbol{\Delta}\mathbf{x} + \boldsymbol{\nu}$$

where we have defined

$$\boldsymbol{\nu} = \boldsymbol{\varepsilon}|_{\mathscr{H}_0} + \mathbf{b}$$

As seen in Sect. 2.1.2.2, this vector can be decomposed in two components by using the decomposition of the space $\mathbb{R}^N$ into two complementary orthogonal subspaces: the column range of $\mathbf{H}$, $Range\{\mathbf{H}\}$, and the null space of $\mathbf{H}^T$, $Null\{\mathbf{H}^T\}$:

$$\mathbb{R}^N = Range\{\mathbf{H}\} \oplus Null\{\mathbf{H}^T\} \tag{3.45}$$

introduced in Sect. 2.1.2.1. More in detail, from Sect. 2.1.2.2, we know that the error vector $\boldsymbol{\nu}$ can be decomposed into its component $\boldsymbol{\nu}_H$ belonging to the range of $\mathbf{H}$, and its component $\boldsymbol{\nu}_\perp$ belonging to the null space of $\mathbf{H}^T$, that is

$$\boldsymbol{\nu} = \boldsymbol{\nu}_H + \boldsymbol{\nu}_\perp \tag{3.46}$$

We know also that the two components can be obtained using the projection matrix $\mathbf{P} = \mathbf{H}(\mathbf{H}^T\mathbf{H})^{-1}\mathbf{H}^T$, that is

$$\begin{cases} \boldsymbol{\nu}_H = \mathbf{P}\boldsymbol{\nu} \\ \boldsymbol{\nu}_\perp = (\mathbf{I} - \mathbf{P})\boldsymbol{\nu} \end{cases} \tag{3.47}$$

From (2.26), we know that $\boldsymbol{\nu}_H$ is the component that determines the position estimation error. On the other hand from (3.12), we know that the observable range residual vector $\hat{\boldsymbol{\varepsilon}}_{rr}$ can be written in the form

$$\hat{\boldsymbol{\varepsilon}}_{rr} = (\mathbf{I} - \mathbf{H}(\mathbf{H}^T\mathbf{H})^{-1}\mathbf{H}^T)\boldsymbol{\nu} = \boldsymbol{\nu}_\perp \tag{3.48}$$

and then the component $\boldsymbol{\nu}_\perp$ is de facto included in the test statistic r_{LS} used to determine whether there is or not a fault. Therefore, $\boldsymbol{\nu}_H$ is the undetectable component of $\boldsymbol{\nu}$, and $\boldsymbol{\nu}_\perp$ contains all the possible information that can be extracted from the error vector $\boldsymbol{\nu}$.

Concerning the solution separation test, it is possible to show that the operation performed by this test is equivalent to a projection of the vector $\boldsymbol{\nu}_\perp$ into a subspace of $Null\{\mathbf{H}^T\}$, performing a sort of "Principal Component Analysis". The estimated position considering all the satellites and the estimated position considering only the reduced subset are

$$\begin{cases} \hat{\boldsymbol{\Delta}\mathbf{x}}|_{\mathscr{H}_0} = (\mathbf{H}^T\mathbf{H})^{-1}\mathbf{H}^T\boldsymbol{\Delta}_\rho \\ \hat{\boldsymbol{\Delta}\mathbf{x}}|_{\mathscr{H}_a} = (\mathbf{H}_R^T\mathbf{H}_R)^{-1}\mathbf{H}_R^T\boldsymbol{\Delta}_\rho \end{cases}$$

The delta solution separation vector is

$$\mathbf{d} = \hat{\boldsymbol{\Delta}\mathbf{x}}|_{\mathscr{H}_0} - \hat{\boldsymbol{\Delta}\mathbf{x}}|_{\mathscr{H}_{a,n}} = ((\mathbf{H}^T\mathbf{H})^{-1}\mathbf{H}^T - (\mathbf{H}_R^T\mathbf{H}_R)^{-1}\mathbf{H}_R^T)\boldsymbol{\Delta}_\rho$$

For simplicity, we define a new matrix $\mathbf{L} = ((\mathbf{H}^T\mathbf{H})^{-1}\mathbf{H}^T - (\mathbf{H}_R^T\mathbf{H}_R)^{-1}\mathbf{H}_R^T)$, and we write the difference vector as

$$\mathbf{d} = \mathbf{L}\boldsymbol{\Delta}_\rho = \mathbf{L}\mathbf{H}\boldsymbol{\Delta}\mathbf{x} + \mathbf{L}\boldsymbol{\nu} = \mathbf{L}\mathbf{H}\boldsymbol{\Delta}\mathbf{x} + \mathbf{L}\boldsymbol{\nu}_H + \mathbf{L}\boldsymbol{\nu}_\perp$$

Since it is easy to show that both $\mathbf{H}\boldsymbol{\Delta}\mathbf{x}$ and $\boldsymbol{\nu}_H$ belong to the null space of $\mathbf{L}$, the difference vector can be simply rewritten as

$$\mathbf{d} = \mathbf{L}\boldsymbol{\nu}_\perp \tag{3.49}$$

It is possible to show that this vector can be also written in the form

$$\mathbf{d} = \frac{\mathbf{G}\mathbf{h}^T}{1-\eta}\nu_{\perp,N_{sat}} \tag{3.50}$$

where $\mathbf{G} = (\mathbf{H}^T\mathbf{H})^{-1}$, and $\eta = \mathbf{h}\mathbf{G}\mathbf{h}^T$.
Proof of (3.54), From (3.29) and (3.44), we can write

$$\mathbf{d} = \hat{\boldsymbol{\Delta}\mathbf{x}}|_{\mathscr{H}_0} - \hat{\boldsymbol{\Delta}\mathbf{x}}|_{\mathscr{H}_{a,n}} = \mathbf{G}\mathbf{h}^T\hat{b} = \mathbf{G}\mathbf{h}^T\nu_{\perp,N_{sat}}(1 + \mathbf{h}(\mathbf{H}_R^T\mathbf{H}_R)^{-1}\mathbf{h}^T)$$

From the Sherman–Morrison formula we have

$$\mathbf{G}_R = (\mathbf{H}^T\mathbf{H} - \mathbf{h}^T\mathbf{h}) = \mathbf{G} + \frac{\mathbf{G}\mathbf{h}^T\mathbf{h}\mathbf{G}}{1 - \mathbf{h}\mathbf{G}\mathbf{h}^T}$$

where $\mathbf{G}_R = (\mathbf{H}_R^T\mathbf{H}_R)^{-1}$. Then, we can write that

$$\mathbf{h}(\mathbf{H}_R^T\mathbf{H}_R)^{-1}\mathbf{h}^T = \mathbf{h}(\mathbf{G} + \frac{\mathbf{G}\mathbf{h}^T\mathbf{h}\mathbf{G}}{1 - \mathbf{h}\mathbf{G}\mathbf{h}^T})\mathbf{h}^T = \eta + \frac{\eta^2}{1-\eta}$$

and consequently

$$\mathbf{h}(\mathbf{H}_R^T\mathbf{H}_R)^{-1}\mathbf{h}^T = \frac{1}{1-\eta}$$

From this last equality, we can write that

$$\mathbf{d} = \mathbf{G}\mathbf{h}^T \nu_{\perp,N_{sat}}(1 + \mathbf{h}(\mathbf{H}_R^T\mathbf{H}_R)^{-1}\mathbf{h}^T) = \mathbf{G}\mathbf{h}^T \frac{\nu_{\perp,N_{sat}}}{1-\eta}$$

□

From (3.49), we see that all the information that can be extracted from the measured data is already present in the range residual vector $\hat{\boldsymbol{\varepsilon}}_{rr} = \boldsymbol{\nu}_{\perp}$ and that the solution separation test can be naturally performed in cascade to the range residual vector test.

An important property that we can extract from the matrix $\mathbf{L}$ is its rank, which is equal to 1. This means that all the information is contained in the first element d_1 of the vector $\mathbf{d}$, whose expression can be obtained from (3.50). At this point, it is better to generalize the results obtained so far and to write the expression of d_1 in the general case of a bias in the ith satellite. Therefore, by adopting a notation similar to the one introduced in Sect. 3.3.3 for the case of multiple faults, we write the first element of the vector $\mathbf{d}$ in the form

$$d_1^{(0\to i)} = \frac{\mathbf{g}\mathbf{h}_i^T}{1-\eta_i}\nu_{\perp,i} \tag{3.51}$$

where $\mathbf{g}$ is the first row of the matrix $\mathbf{G}$.

It is evident that $d_1^{(0\to i)}$ can be used as a test statistic to perform the SS test instead of the classical method based on the difference vector $\mathbf{d} = \hat{\boldsymbol{\Delta}}\mathbf{x}|_{\mathscr{H}_0} - \hat{\boldsymbol{\Delta}}\mathbf{x}|_{\mathscr{H}_{a,i}}$. The advantage of this equivalent formulation is that it has a lower computational cost, since it does not require the computation of a new position for each hypothesis under test. Moreover, the term $\mathbf{g}\mathbf{h}_i^T$ that is included in the computation of the position with all the satellites in view therefore has not to be computed again for each hypothesis $\mathscr{H}_{a,i}$.

3.5.1 *Efficient Implementation in Single Fault Case*

In this section, we present the average running time per epochs of the SS test in the single fault case using the classical implementation [12–15] of the SS algorithm or the one presented in the previous section, showing the drastic improvement of performance using the presented scheme. In the case of a single fault case, it is possible to further simplify the decision test in the form

$$\frac{d_1^{(0\to i)}}{T_1^{(0\to i)}} = \frac{\nu_{\perp,i}}{\sqrt{1-\eta_i}} \tag{3.52}$$

where the expression of $T_1^{(0\to i)}$, derived from (3.39), is

$$T_1^{(0\to i)} = \sqrt{(\mathbf{L}^{(0\to i)}\mathbf{L}^{(0\to i)^T})_{1,1}} \tag{3.53}$$

Notice that the test is actually performed on the absolute value of (3.52).

To prove (3.52), we have first to prove that

$$\mathbf{L}^{(0\to i)}\mathbf{L}^{(0\to i)^T} = \frac{\mathbf{G}\mathbf{h}_i^T\mathbf{h}_i\mathbf{G}}{1-\eta_i} \tag{3.54}$$

This expression combined with (3.53) leads to write

$$T_1^{(0\to i)} = \sqrt{(\mathbf{L}^{(0\to i)}\mathbf{L}^{(0\to i)^T})_{1,1}} = \frac{\mathbf{g}\mathbf{h}_i^T}{\sqrt{1-\eta_i}} \tag{3.55}$$

which combined with (3.51) leads to (3.52).

Proof of (3.54), From (3.40), and by recalling that $\mathbf{G}^T = \mathbf{G}$, we derive

$$\mathbf{L}^{(0\to i)}\mathbf{L}^{(0\to i)^T} = ((\mathbf{G}\mathbf{H}^T - \mathbf{G}_{(T\setminus i)}\mathbf{H}^T_{(T\setminus i)})(\mathbf{H}\mathbf{G} - \mathbf{H}_{(T\setminus i)}\mathbf{G}_{(T\setminus i)})$$

that can be expanded as

$$\begin{aligned}
&\mathbf{G}\mathbf{H}^T\mathbf{H}\mathbf{G} - \mathbf{G}_{(T\setminus i)}\mathbf{H}^T_{(T\setminus i)}\mathbf{H}\mathbf{G}\\
&- \mathbf{G}\mathbf{H}^T\mathbf{H}_{(T\setminus i)}\mathbf{G}_{(T\setminus i)} - \mathbf{G}_{(T\setminus i)}\mathbf{H}^T_{(T\setminus i)}\mathbf{H}_{(T\setminus i)}\mathbf{G}_{(T\setminus i)} =\\
&\mathbf{G} - \mathbf{G} - \mathbf{G} + \mathbf{G}_{(T\setminus i)} = \mathbf{G}_{(T\setminus i)} - \mathbf{G}
\end{aligned}$$

since $\mathbf{H}^T_{(T\setminus i)}\mathbf{H} = \mathbf{H}^T_{(T\setminus i)}\mathbf{H}_{(T\setminus i)}$. Moreover, from the Sherman–Morrison formula, we have

$$\mathbf{G}_{(T\setminus i)} = (\mathbf{H}^T\mathbf{H} - \mathbf{h}_i^T\mathbf{h}_i) = \mathbf{G} + \frac{\mathbf{G}\mathbf{h}_i^T\mathbf{h}_i\mathbf{G}}{1-\mathbf{h}_i\mathbf{G}\mathbf{h}_i^T}$$

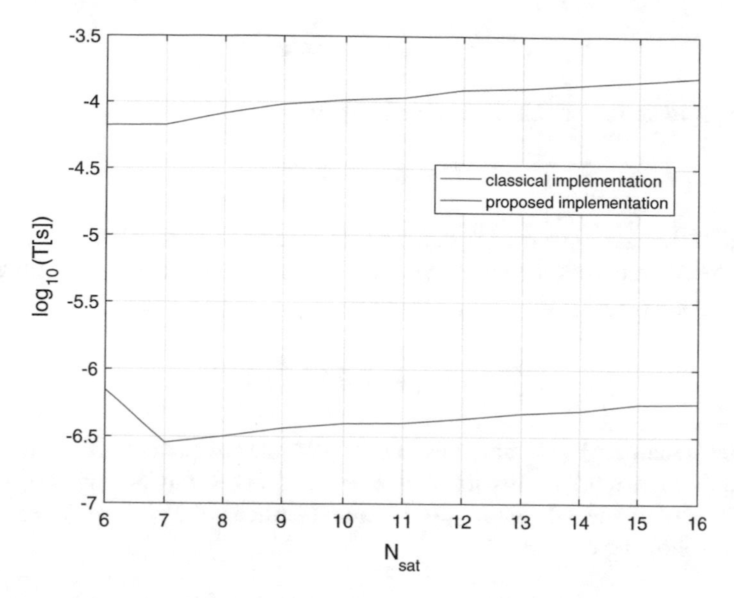

Fig. 3.2 Single fault case comparison, on the abscissa axis the number of in view satellite N_{sat}, on the ordinate axis the average running time per epoch (normalized with respect to $1s$ and in log scale)

and then, we can write that

$$\mathbf{L}^{(0\to i)}\mathbf{L}^{(0\to i)T} = \mathbf{G}_{(T\setminus i)} - \mathbf{G} = \frac{\mathbf{G}\mathbf{h}_i^T\mathbf{h}_i\mathbf{G}}{1-\mathbf{h}_i\mathbf{G}\mathbf{h}_i^T} \tag{3.56}$$

□

In Fig. 3.2, it is reported the average running time per epoch versus a different number of in view satellites, where the classical and proposed algorithms are compared and it is evident the increase in performance.

3.5.2 Efficient Implementation in Double Fault Case

In this section, we develop the theory necessary to understand how the low-complexity algorithm can be implemented in the case of a maximum of simultaneous faults equal to 2. Usually, in avionic environment, two simultaneous faults are rare enough to avoid testing three or more simultaneous faults [13].

Before proceeding in the description, it can be useful to introduce and formalize some quantities.

Suppose that we want to compute the difference between the all-in-view position and the position excluding satellite i and satellite j. We define this difference as

$$\mathbf{d}^{(0\to i,j)} = \hat{\boldsymbol{\Delta}\mathbf{x}}^{(T)} - \hat{\boldsymbol{\Delta}\mathbf{x}}^{(T\setminus i,j)} \tag{3.57}$$

It is simple to show that this difference is equal to

$$\mathbf{d}^{(0\to i,j)} = \mathbf{d}^{(0\to i)} + \mathbf{d}^{(i\to i,j)} \tag{3.58}$$

where $\mathbf{d}^{(0\to i)} = \hat{\boldsymbol{\Delta}\mathbf{x}}^{(T)} - \hat{\boldsymbol{\Delta}\mathbf{x}}^{(T\setminus i)}$ and $\mathbf{d}^{(i\to i,j)} = \hat{\boldsymbol{\Delta}\mathbf{x}}^{(T\setminus i)} - \hat{\boldsymbol{\Delta}\mathbf{x}}^{(T\setminus i,j)}$.

For the derivation of the low-complexity algorithm, it is useful to introduce the reduced residual vector $\boldsymbol{\nu}_{\perp}^{(T\setminus i)}$ of dimension $(N_{sat} - i) \times 1$, defined as

$$\boldsymbol{\nu}_{\perp}^{(T\setminus i)} = \{\boldsymbol{\Delta}_{\rho} - \mathbf{H}\hat{\boldsymbol{\Delta}\mathbf{x}}^{(T\setminus i)}\} \downarrow_{N_{sat}}^{i+1} \tag{3.59}$$

where the notation $\{\cdot\} \downarrow_{N_{sat}}^{i+1}$ has to be intended as: the components of $\cdot$ from element $i+1$ up to element N_{sat}. It will be clear looking at the implementation why it is sufficient to compute only the $N_{sat} - i$ elements instead of $N_{sat} - 1$. We can readily show the following equivalence

$$\boldsymbol{\nu}_{\perp}^{(T\setminus i)} = \left\{\boldsymbol{\nu}_{\perp}^{(T\setminus 0)} + \frac{\mathbf{H}\mathbf{G}\mathbf{h}_i^T}{1 - \mathbf{h}_i\mathbf{G}\mathbf{h}_i^T}\nu_{\perp,i}^{(T\setminus 0)}\right\} \downarrow_{N_{sat}}^{i+1} \tag{3.60}$$

where, with the notations of the previous sections, $\boldsymbol{\nu}_{\perp}^{(T\setminus 0)} = \boldsymbol{\nu}_{\perp}$. Notice that in (3.60), all the elements are already known and then the new residual vector can be efficiently computed.

Proof of (3.60), We start with the consideration that, since

$$\boldsymbol{\nu}_{\perp}^{(T\setminus i)} = \{\boldsymbol{\Delta}_{\rho} - \mathbf{H}\hat{\boldsymbol{\Delta}\mathbf{x}}^{(T\setminus i)}\} \downarrow_{N_{sat}}^{i+1}$$

and

$$\mathbf{H}\hat{\boldsymbol{\Delta}\mathbf{x}}^{(T\setminus i)} = \mathbf{H}(\hat{\boldsymbol{\Delta}\mathbf{x}}^{(T\setminus 0)} - \mathbf{d}^{(0\to i)})$$

we can write that

$$\boldsymbol{\Delta}_{\rho} - \mathbf{H}\hat{\boldsymbol{\Delta}\mathbf{x}}^{(T\setminus i)} = \boldsymbol{\Delta}_{\rho} - \mathbf{H}(\hat{\boldsymbol{\Delta}\mathbf{x}}^{(T\setminus 0)} - \mathbf{d}^{(0\to i)})$$

Since

$$\boldsymbol{\Delta}_{\rho} - \mathbf{H}\hat{\boldsymbol{\Delta}\mathbf{x}}^{(T\setminus 0)} = \boldsymbol{\nu}_{\perp}^{(T\setminus 0)}$$

and, from (3.51),

$$\mathbf{d}^{(0\to i)} = \frac{\mathbf{G}\mathbf{h}_i^T}{1 - \eta_i}\nu_{\perp,i}$$

we can derive that (3.60) is correct. □

Thanks to the introduction of the quantity $\boldsymbol{\nu}_{\perp}^{(T\setminus i)}$ it is possible to write

$$\mathbf{d}^{(i\rightarrow i,j)} = \frac{\mathbf{G}_{(T\setminus i)}\mathbf{h}_j^T}{1-\mathbf{h}_j\mathbf{G}_{(T\setminus i)}\mathbf{h}_j^T}\nu_{\perp,j}^{(T\setminus i)} \tag{3.61}$$

In Sect. 3.5.1, we have seen that in the case of a single fault the only quantity that has to be computed is (3.52). Now in the case of two simultaneous faults, we have to test the following quantities

$$\frac{d_p^{(0\rightarrow i,j)}}{T_p^{(0\rightarrow i,j)}}, \quad p = 1, 2, 3 \tag{3.62}$$

The first step for the simplification of (3.62) can be derived thanks to the following property

$$T_p^{(0\rightarrow i,j)} = \sqrt{T_p^{(0\rightarrow i)2} + T_p^{(i\rightarrow i,j)2}} \tag{3.63}$$

Proof of (3.63), From (3.40)

$$\mathbf{L}^{(0\rightarrow i,j)} = ((\mathbf{H}^T\mathbf{H})^{-1}\mathbf{H}^T - (\mathbf{H}_{(T\setminus i),j}^T\mathbf{H}_{(T\setminus i),j})^{-1}\mathbf{H}_{(T\setminus i),j}^T)$$

that can be rewritten as

$$\begin{aligned}((\mathbf{H}^T\mathbf{H})^{-1}\mathbf{H}^T - (\mathbf{H}_{(T\setminus i)}^T\mathbf{H}_{(T\setminus i)})^{-1}\mathbf{H}_{(T\setminus i)}^T +\\ (\mathbf{H}_{(T\setminus i)}^T\mathbf{H}_{(T\setminus i)})^{-1}\mathbf{H}_{(T\setminus i)}^T - (\mathbf{H}_{(T\setminus i),j}^T\mathbf{H}_{(T\setminus i),j})^{-1}\mathbf{H}_{(T\setminus i),j}^T)\end{aligned}$$

we notice that

$$\mathbf{L}^{(0\rightarrow i,j)} = \mathbf{L}^{(0\rightarrow i)} + \mathbf{L}^{(i\rightarrow i,j)}$$

Moreover since

$$\begin{aligned}\mathbf{L}^{(0\rightarrow i)}\mathbf{L}^{(i\rightarrow i,j)T} =\\ ((\mathbf{H}^T\mathbf{H})^{-1}\mathbf{H}^T - (\mathbf{H}_{(T\setminus i)}^T\mathbf{H}_{(T\setminus i)})^{-1}\mathbf{H}_{(T\setminus i)}^T)\\ (\mathbf{H}_{(T\setminus i)}(\mathbf{H}_{(T\setminus i)}^T\mathbf{H}_{(T\setminus i)})^{-1} - \mathbf{H}_{(T\setminus i),j}(\mathbf{H}_{(T\setminus i),j}^T\mathbf{H}_{(T\setminus i),j})^{-1}) =\\ \mathbf{0}\end{aligned}$$

the final conclusion is that

$$\mathbf{L}^{(0\rightarrow i,j)T}\mathbf{L}^{(0\rightarrow i,j)} = \mathbf{L}^{(0\rightarrow i)}\mathbf{L}^{(0\rightarrow i)T} + \mathbf{L}^{(0\rightarrow i,j)}\mathbf{L}^{(0\rightarrow i,j)T}$$

and in particular

$$T_p^{(0\rightarrow i,j)} = \sqrt{T_p^{(0\rightarrow i)2} + T_p^{(i\rightarrow i,j)2}}$$

□

It is then possible to rewrite (3.62) as

$$\frac{d_p^{(0\rightarrow i,j)}}{T_p^{(0\rightarrow i,j)}} = \frac{d_p^{0\rightarrow i}}{T_p^{(0\rightarrow i)}\sqrt{1+\left(\frac{T_p^{(i\rightarrow i,j)}}{T_p^{(0\rightarrow i)}}\right)^2}} + \frac{d_p^{i\rightarrow i,j}}{T_p^{(i\rightarrow i,j)}\sqrt{1+\left(\frac{T_p^{(i\rightarrow i,j)}}{T_p^{(0\rightarrow i)}}\right)^{-2}}} \tag{3.64}$$

where the two quantities $\frac{d_p^{0\rightarrow i}}{T_p^{(0\rightarrow i)}}$ and $\frac{d_p^{i\rightarrow i,j}}{T_p^{(i\rightarrow i,j)}}$ can be written as in (3.52) in the form $\frac{\nu_{\perp,i}}{\sqrt{1-\eta_i}}$ and $\frac{\nu_{\perp,j}^{(T\backslash i)}}{\sqrt{1-\eta_j^i}}$, independently from the value of p. Finally, (3.64) can be rewritten as

$$\frac{d_p^{(0\rightarrow i,j)}}{T_p^{(0\rightarrow i,j)}} = \frac{\frac{\nu_{\perp,i}}{\sqrt{1-\eta_i}}}{\sqrt{1+\left(\frac{T_p^{(i\rightarrow i,j)}}{T_p^{(0\rightarrow i)}}\right)^2}} + \frac{\frac{\nu_{\perp,j}^{(T\backslash i)}}{\sqrt{1-\eta_j^i}}}{\sqrt{1+\left(\frac{T_p^{(i\rightarrow i,j)}}{T_p^{(0\rightarrow i)}}\right)^{-2}}} \tag{3.65}$$

where from (3.56)

$$\eta_j^i = \mathbf{h}_j \mathbf{G}_{(T\backslash i)} \mathbf{h}_j^T = \mathbf{h}_j \mathbf{G} \mathbf{h}_j^T + \frac{(\mathbf{h}_j \mathbf{G} \mathbf{h}_i^T)^2}{1-\eta_i} \tag{3.66}$$

and

$$\frac{T_p^{(i\rightarrow i,j)}}{T_p^{(0\rightarrow i)}} = \frac{\mathbf{g}_p \mathbf{h}_j^T}{\mathbf{g}_p \mathbf{h}_i^T} + \mathbf{h}_j \mathbf{G} \mathbf{h}_i^T \tag{3.67}$$

We can finally present the pseudocode in Procedure 1 for the implementation of the algorithm. Figure 3.3 shows the performance for the case of 1 and 2 simultaneous faults for the proposed and classical algorithms, and also in this case, we have a performance gain.

In general, extending the SS algorithm to the case of simultaneous multiple faults, it is possible to generalize the derived results and recursively efficiently compute any difference vector $\mathbf{d}^{(S_i)}$. Figure 3.4 represents the tree structure on which the SS tests can be efficiently implemented, in the case of a constellation with $N_{sat} = 4$ and a maximum number of simultaneous faults equal to 3. Obviously such a system cannot exist but the figure is useful for explaining the concept.

Procedure 1 Two Simultaneous faults solution separation tests

Input: $\mathbf{HGH}^T, \mathbf{GH}^T$,

Output: $\frac{d_i^{(0\to i)}}{T_1^{(0\to i)}}, \frac{d_p^{(0\to i,j)}}{T_p^{(0\to i,j)}}$

1: **for** $(i = 1 : N_{sat})$ **do**

2: $\quad \frac{d_1^{0\to i}}{T_1^{(0\to i)}} = \frac{\nu_{\perp,i}}{\sqrt{1-\eta_i}}$

3: $\quad \boldsymbol{\nu}_\perp^{(T\backslash i)} = \{\boldsymbol{\nu}_\perp^{(T\backslash 0)} + \frac{\mathbf{HGh}_i^T}{1-\mathbf{h}_i\mathbf{Gh}_i^T}\nu_{\perp,i}^{(T\backslash 0)}\} \downarrow_{N_{sat}}^{i+1}$

4: $\quad$ **for** $(j = i : N_{sat})$ **do**

5: $\quad\quad$ **for** $(p = 1 : 3)$ **do**

6:

$$\frac{d_p^{(0\to i,j)}}{T_p^{(0\to i,j)}} = \frac{\frac{\nu_{\perp,i}}{\sqrt{1-\eta_i}}}{\sqrt{1+\left(\frac{\mathbf{g}_p\mathbf{h}_j^T}{\mathbf{g}_p\mathbf{h}_i^T}+\mathbf{h}_j\mathbf{Gh}_i^T\right)^2}} + \frac{\frac{\nu_{\perp,j}^{(T\backslash i)}}{\sqrt{1-\eta_j^i}}}{\sqrt{1+\left(\frac{\mathbf{g}_p\mathbf{h}_j^T}{\mathbf{g}_p\mathbf{h}_i^T}+\mathbf{h}_j\mathbf{Gh}_i^T\right)^{-2}}} \tag{3.68}$$

7: $\quad\quad$ **end for**

8: $\quad$ **end for**

9: **end for**

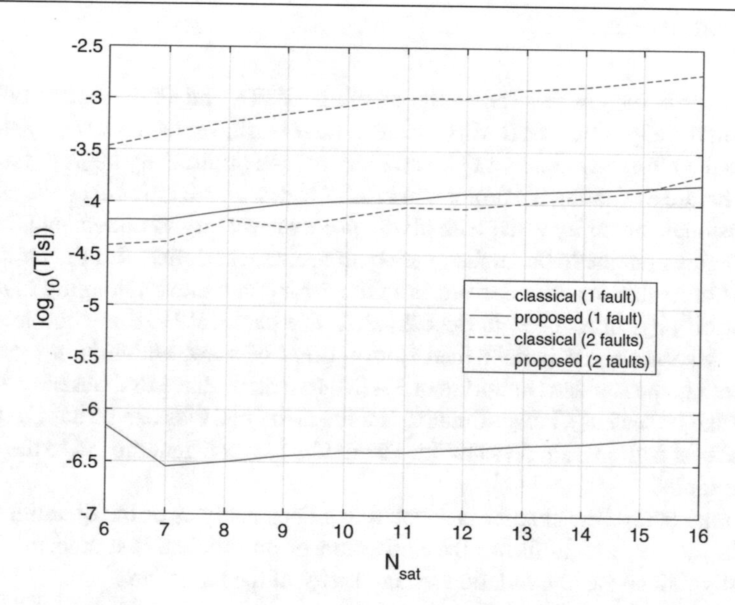

Fig. 3.3 Single fault and double fault case comparison, on the abscissa axis the number of in view satellite N_{sat}, on the ordinate axis the average running time per epoch (normalized with respect to 1s and in log scale)

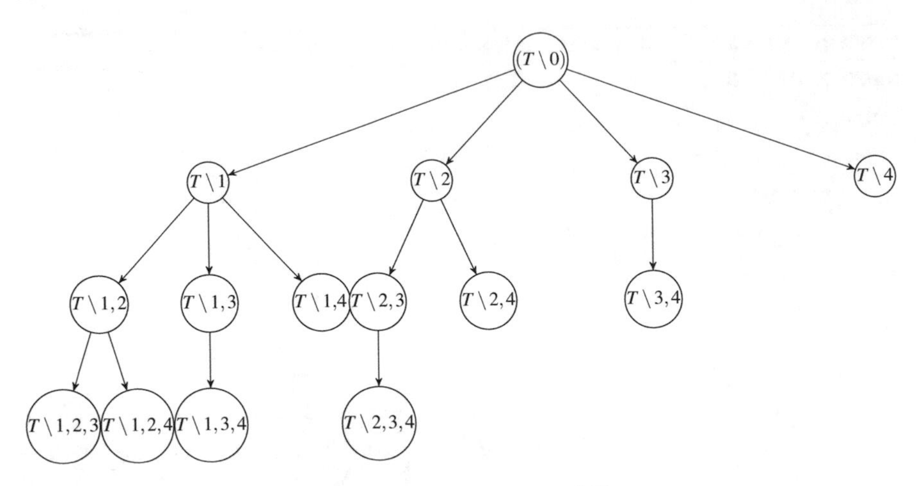

Fig. 3.4 Tree structure useful for the efficient computation of SS

3.6 Conclusions

This chapter is mainly devoted to the methods of FD, and FDE, generally implemented in the algorithms of RAIM systems. First of all, we have seen why the tests performed by the algorithms of FD, and FDE, are the preliminary steps to be implemented before evaluating the protection level. Then, we have introduced the classical range residual and parity tests, with all the proofs necessary to understand that they are two equivalent methods. In fact, they both use, as test statistic for the fault detection, an observable quantity derived from the error component belonging to the null space of $\mathbf{H}^T$, $\epsilon_{\perp}$, introduced in Sect. 2.1.2.2. The classical FDE method known as solution separation test is described from a point of view, which is not generally considered in the standard literature of RAIM. In fact, we have first observed that the method can be seen as a bias estimator, and the estimated bias can be used to test the presence of a fault in a single satellite. The method is then generalized to the case of multiple faults.

The final part of the chapter is devoted to a deep analysis of the quantities introduced in the SS, which allows the derivation of an efficient test statistic, able to drastically reduce the computational complexity of the algorithm.

References

1. Kaplan E, Hegarty C (2006) Undestanding GPS: principles and applications, 2nd edn. Artech House
2. DC Radio Technical Commission for Aeronautics (1991) Washington. Minimum operational performance standards for airborne supplemental navigation equipment using global positioning system (GPS). RTCA/DO-208

3. Grover Brown R (1992) A baseline RAIM scheme and a note on the equivalence of three RAIM methods. NAVIGATION, J Inst Navig 39(3):301–316
4. Parkinson BW, Axelrad P (1988) Autonomous GPS integrity monitoring using the pseudorange residual. NAVIGATION: J Inst Navig (USA) 35(2):255
5. Gleason S (2009) GNSS applications and methods - GNSS technology and applications. Artech House
6. Liu J, Lu M, Feng Z, Wang J (2005) GPS RAIM: statistics based improvement on the calculation of threshold and horizontal protection radius. In: International Symposium on GPS/GNSS, Hong Kong, pp 8–10
7. Sturza MA, Brown AK (1990) Comparison of fixed and variable threshold RAIM algorithms. In: Proceedings of the 3rd International Technical Meeting of the Satellite Division of The Institute of Navigation (ION GPS 1990), Colorado Spring, CO, pp 437–443
8. Sturza MA (1988–1989) Navigation system integrity monitoring using redundant measurements. NAVIGATION, J Inst Navig 35(4):483–502
9. Dahlquist G, Björck Å (1974) Numerical methods. Prentice-Hall, Englewood Cliffs
10. Kay SM (1998) Fundamentals of statistical signal processing: Detection theory, vol II. Prentice-Hall, Upper Saddle River
11. Sherman J, Morrison WJ (1950) Adjustment of an inverse matrix corresponding to a change in one element of a given matrix. Ann Math Stat 21(1):124–127
12. Blanch J, Walker T, Enge P, Lee Y, Pervan B, Rippl M, Spletter A, Kropp V (2015) Baseline advanced RAIM user algorithm and possible improvements. IEEE Trans Aerosp Electr Syst 51(1):713–732
13. Milestone 3 report. Technical report, GPS-Galileo Working Group C ARAIM Technical Subgroup, 26 February 2016
14. Joerger M, Pervan B (2014) Solution separation and Chi-Squared ARAIM for fault detection and exclusion. In: 2014 IEEE/ION Position, Location and Navigation Symposium - PLANS 2014, pp 294–307
15. Joerger M, Stevanovic S, Chan FC, Langel S, Pervan B (2013) Integrity risk and continuity risk for fault detection and exclusion using solution separation ARAIM . In: Proceedings of the 26th International Technical Meeting of The Satellite Division of the Institute of Navigation (ION GNSS), Nashville, Tennessee, pp 2702–2722

Chapter 4
Evaluation of the Confidence Intervals

Letizia Lo Presti and Giulio Franzese

Abstract This chapter describes the methods of evaluation of the confidence interval of the estimated position (the PL), taking into account the remaining errors, which may be present in the measured pseudoranges, after the application of the FD and FDE algorithms. We have seen in the previous chapter that these algorithms are able to exclude undesired situations. The position is generally considered unavailable if the FD detects unacceptable faults, and then in this case it is not necessary to evaluate the confidence interval, since there is not an estimated position. If the faulty satellites are excluded by the FDE algorithm, the position is accepted, and the confidence interval has to be evaluated. However undetectable errors still remain in the estimated position and they have to be taken into account in the PL computation.

4.1 Confidence Interval in the Case of a Single Fault

In Sect. 3.1 we have seen how to compute the confidence interval of the estimated position in the simple case of position coordinates affected by only AWGN. The confidence interval we are interested in is the PL defined in Sect. 3.1. In theory, if all the faulty satellites are excluded by the FDE algorithm we can simply use (3.1) to compute the PL, as described in Sect. 3.1.1. In practice it may happen that the test statistic fails to detect a fault, and this event has to be taken into account in the evaluation of the confidence interval.

In the previous chapter we have introduced two equivalent test statistics r_{LS} and r_p to check the integrity of the estimated position. Since our interest is now to quantify in some way the Position Error (PE), in this section we examine the relationship

---

L. Lo Presti (✉) · G. Franzese
Politecnico di Torino, c.so Duca degli Abruzzi n.24, Torino, Italy
e-mail: letizia.lopresti@polito.it

G. Franzese
e-mail: giulio.franzese@polito.it

© Springer International Publishing AG, part of Springer Nature 2018

L. Lo Presti and S. Sabina (eds.), *GNSS for Rail Transportation*, PoliTO Springer Series, https://doi.org/10.1007/978-3-319-79084-8_4

between the observable r_{LS} and the unobservable PE. To introduce this topic we start with the case of a single satellite fault, and we assume that the fault generates a deterministic unknown bias in the pseudorange of the faulty satellite. Finally from this analysis we derive a method to include an undetected bias in the evaluation of the PL. Notice that the PL can be defined for both the Vertical component VPL and for the Horizontal component HPL of the estimated position; in this chapter we mainly focus for simplicity on the VPL but similar methods are described in literature for the computation of HPL.

4.1.1 Alert Limit, Integrity Risk, and Protection Level

The PL is generally computed to decide if a position measurement is reliable or not. To do this, the concept of Alert Limit (AL) limit has to be introduced.

The horizontal (respectively vertical) *alert limit* is the maximum allowable horizontal (respectively vertical) position error beyond which a GNSS system should be declared not able to provide the position information for the intended operations. In fact if the position error exceeds the alert limit, and the positioning system is not able to recognize the occurrence of this event, the user experiences a risk if he uses the estimated position for his application. For this reason in the GNSS community the probability that, at any moment, the position error exceeds the AL is called integrity risk. If such a condition is detected a Loss Of Integrity (LOI) message is sent to the user. Notice that in many applications the definition of integrity risk is related to a time interval, where N measurements are performed. In this case it is necessary to distinguish between the global IR (in the whole time interval) and the punctual IR (in a single measurement). The relationship between the global and punctual IRs depends on the exact definition of the global IR, and on the error models adopted for each specific application. For example, in aviation applications, the punctual IR is obtained by dividing the global IR by N. Details on the method to introduce the punctual IR can be found in [1, 2].

From now on we use the term IR for the punctual case, unless differently specified.

Since the position error is not observable, it is not possible to verify if the threshold AL has been passed or not at each measurement. For this reason the observable quantity PL is used for the comparison against the AL. In Sect. 3.1.1 we have seen that this quantity tries to quantify the maximum absolute error. Since this error is theoretically infinite, the PL can only identifies a “quasi-sure” error range. To quantify how “sure” is this range, a target IR is fixed and the PL becomes the confidence interval of the position error related to this target IR.

Figure 4.1 shows examples of situations involving PL and PE in the case of a PE with a Gaussian PDF.

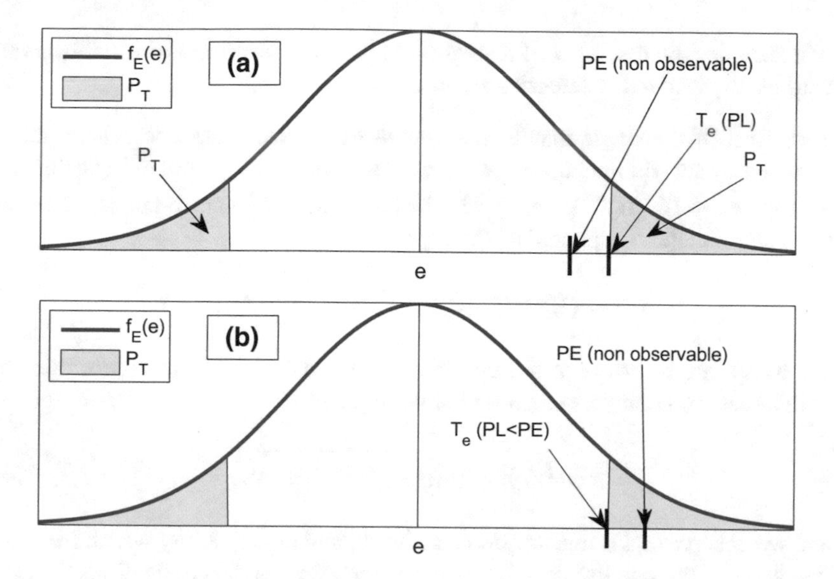

Fig. 4.1 Examples of values of PL and PE in the case of a position error with a Gaussian PDF

4.1.2 The Effect of Bias in a Single Satellite

We start analyzing the effect of a bias in the measured pseudoranges on the estimated position. In this case the pseudorange errors can be modeled as

$$\varepsilon|_{\mathscr{H}_{a,i},i} = b + \varepsilon_{\mathscr{H}_0,i} \tag{4.1}$$

where b is a bias affecting the i-th satellite. If only a single satellite is biased we can introduce the vector $\boldsymbol{\varepsilon}_{\mathscr{H}_{a,i}} = \boldsymbol{\varepsilon}_{\mathscr{H}_0} + \mathbf{b}$, which characterizes this situation.

$$\mathbf{b} = [0, \ldots, b, 0, \ldots, 0]^T \tag{4.2}$$

The notation used here is the same notation introduced in Sect. 3.3.1.

The hypothesis that only a single satellite is affected by a bias is generally adopted in aviation applications, where the probability of simultaneous faulty pseudoranges is assumed to be negligible. This assumption is realistic, especially when a user is assisted by a GBAS, whose monitoring station picks up and isolates any faulty satellite within a relatively short period of time. In other application scenarios this hypothesis is not realistic, and the analysis of the effects of bias is more complex. In this section we consider only the case of a single bias, and we describe the methods of evaluating the confidence interval only for this case. This method can be seen as the starting point to introduce the more general case of multiple faults, described in Sect. 4.2.3.

We analyse now the effect of a single bias in the observable and non-observable quantities involved in the integrity algorithms.

- The effect of the single bias in the user position error (non-observable) can be obtained by substituting $\boldsymbol{\Delta\rho}$ of (2.11) with the vector $\boldsymbol{\varepsilon}|_{\mathscr{H}_{a,i}}$, obtaining an error of the type $\hat{\boldsymbol{\delta}}\mathbf{x} = (\mathbf{H}^T\mathbf{H})^{-1}\mathbf{H}^T\boldsymbol{\varepsilon}|_{\mathscr{H}_{a,i}}$. In the ideal case of no noise this error reduces to a deterministic component of the type

$$\hat{\boldsymbol{\delta}}\mathbf{x}_{b,i} = (\mathbf{H}^T\mathbf{H})^{-1}\mathbf{H}^T\mathbf{b}_i = b[A_{1i}, A_{2i}, A_{3i}, A_{4i}]^T$$

where A_{ji} are elements of the matrix $\mathbf{A} = (\mathbf{H}^T\mathbf{H})^{-1}\mathbf{H}^T$. To represent this error with a scalar quantity we consider the radial error

$$||\hat{\boldsymbol{\delta}}\mathbf{x}_R|| = |b|\sqrt{A_{1i}^2 + A_{2i}^2 + A_{3i}^2} \tag{4.3}$$

and we observe that it linearly depends on $|b|$ and on a quantity which takes take into account the satellite-user geometry. Depending on the applications the radial error can be restricted to the horizontal plane or to the vertical coordinate. In this section we consider only the 3-D radial error $||\hat{\boldsymbol{\delta}}\mathbf{x}_R||^2$. Notice that (4.3) can be also rewritten as

$$||\hat{\boldsymbol{\delta}}\mathbf{x}_R|| = |b|\sqrt{\mathbf{h}_i\mathbf{G}_{1:3}^2\mathbf{h}_i^T} = |b_i|\sqrt{C_{ii}} \tag{4.4}$$

where we have defined the matrix $\mathbf{C} = \mathbf{H}\mathbf{G}_{1:3}^2\mathbf{H}^T$ and where $\mathbf{G}_{1:3}^2$ is the matrix $\mathbf{G} = (\mathbf{H}^T\mathbf{H})^{-1}$ reduced to the first three rows and columns. The extension to other radial errors is left to the reader.

- The effect of the bias in the observable test statistic ε_{SSE}, defined in (3.8), can be obtained by writing

$$\varepsilon_{\text{SSE}} = [(\mathbf{I}_n - \mathbf{P})\boldsymbol{\varepsilon}|_{\mathscr{H}_{a,i}}]^T(\mathbf{I}_n - \mathbf{P})\boldsymbol{\varepsilon}|_{\mathscr{H}_{a,i}} \tag{4.5}$$

obtained by using (3.12). Thanks to the particular structure of the matrix $\mathbf{P} = \mathbf{H}(\mathbf{H}^T\mathbf{H})^{-1}\mathbf{H}^T$ it is possible to show that

$$\varepsilon_{\text{SSE}} = \boldsymbol{\varepsilon}|_{\mathscr{H}_{a,i}}^T(\mathbf{I}_n - \mathbf{P})\boldsymbol{\varepsilon}|_{\mathscr{H}_{a,i}} \tag{4.6}$$

We know from Sect. 3.2.2.1 that ε_{SSE} has a chi-squared distribution, when no bias is present. In the presence of bias the distribution becomes a non-central chi-squared with $N_{sat} - 4$ degrees of freedom, whose non-centrality parameter λ_i has to be determined. To do this we put $\boldsymbol{\varepsilon} = 0$ in ε_{SSE}, and we extract the deterministic component

$$\varepsilon_{\text{SSE},b,i} = \mathbf{b}^T(\mathbf{I}_n - \mathbf{P})\mathbf{b} = b^2(1 - P_{ii}) \tag{4.7}$$

from which the effect of the deterministic bias in the test statistic r_{LS} can be written as:

$$r_{LS,b,i} = \sqrt{\frac{\varepsilon_{\text{SSE},b,i}}{N_{sat} - 4}} = \frac{|b|\sqrt{(1 - P_{ii})}}{\sqrt{N_{sat} - 4}}$$

a parameter which linearly depends on b and on a quantity which takes take into account the satellite-user geometry.

The non-centrality parameter can be obtained from (4.7), and has the expression

$$\lambda_i = \frac{b^2(1 - P_{ii})}{\sigma_\varepsilon^2} \tag{4.8}$$

valid when all the noise components ε_i have the same variance σ_ε^2.

As a consequence the PDF of $r_{LS,b,i}$ is a chi distribution with a non-centrality parameter

$$\lambda_{r,i} = \sqrt{\frac{\lambda_i}{N_{sat} - 4}} \tag{4.9}$$

4.1.2.1 Simulation Results

We present here some Monte Carlo computer simulation results which show the radial error versus the test statistic r_{LS}, in the presence of noise and a bias in a single satellite. The simulation scenario has been obtained by considering the GPS satellites in view in a single epoch at the location with ECEF coordinates (4472417.109, 601428.9932, 4492692.896). The number of acquired satellites was $N_{sat} = 9$, and the satellite positions have been evaluated by demodulating the navigation message.

In order to assess the performance of the test statistic r_{LS} in nominal and biased conditions, instead of the measured pseudoranges, the geometrical ranges $\rho_{geo,i}$ between the user position and the N_{sat} satellites have been used in the simulation experiments. For the nominal condition the simulated pseudoranges are

$$\rho_{sim,i} = \rho_{geo,i} + \varepsilon_i + b_{clock}$$

where ε_i is a Gaussian random variable with zero mean and variance σ_ε^2, and b_{clock} is a constant which simulates the synchronization error. For the faulty condition the simulated pseudoranges are

$$\rho_{sim,i} = \rho_{geo,i} + \varepsilon_i + b_{clock} + b$$

where b is a bias which simulates a faulty condition.

In the experiments many trials have been done, with $\sigma_\varepsilon = 5$m, and $b \neq 0$ in a single satellite at a time. The simulation results are shown in the figures from 4.2 to 4.5, which show the radial error versus the test statistic. The plots highlight the effects of the random error, the bias, and the geometry. Figures 4.2 and 4.3 show the results for two different values of bias b (respectively $b = 50\,\text{m}$ and $b = 30\,\text{m}$) when the bias affects the pseudorange of the GPS satellite with ID = 7. The marker "o" is used for the results in nominal condition, while the marker "x" is related to

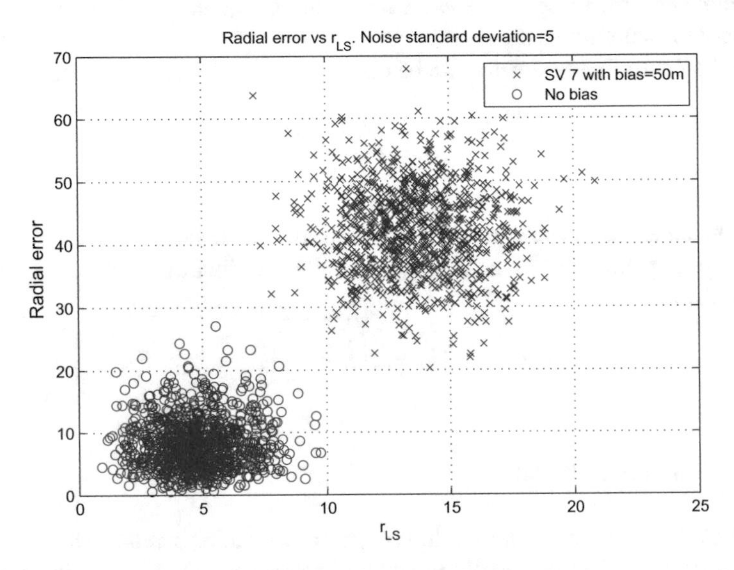

Fig. 4.2 Radial error versus test statistic. Comparison between nominal condition (marker o) and bias in the satellite with ID = 7 (marker x). The bias is 50 m

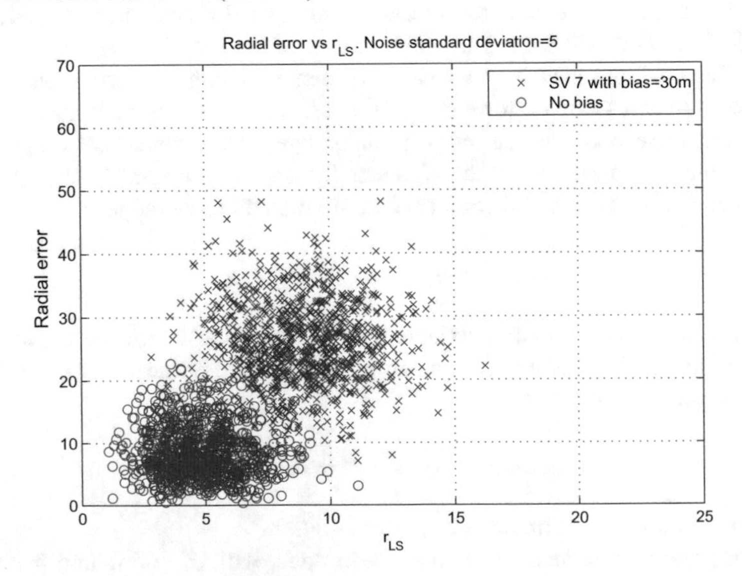

Fig. 4.3 Radial error versus test statistic. Comparison between nominal condition (marker o) and bias in the satellite with ID = 7 (marker x). The bias is 30 m

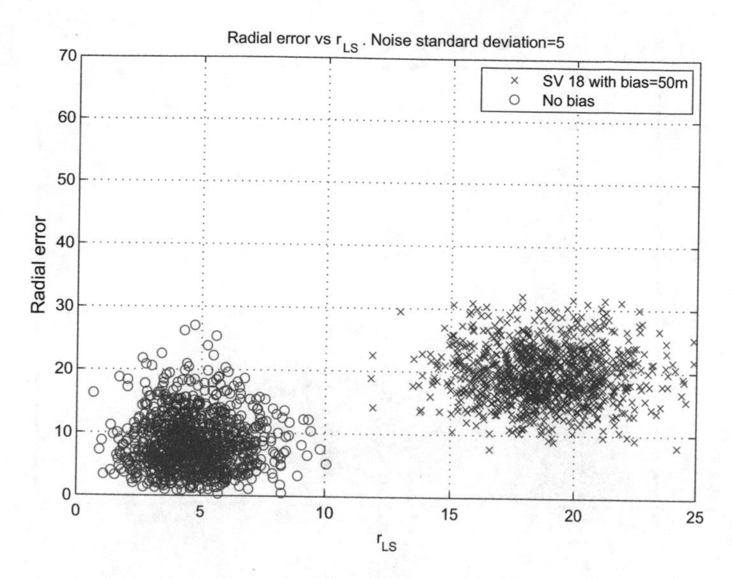

Fig. 4.4 Radial error versus test statistic. Comparison between nominal condition (marker o) and bias in the satellite with ID = 18 (marker x). The bias is 50 m

the faulty condition. In Figs. 4.4 and 4.5 the same analysis has been performed, but with the bias affecting the GPS satellite with ID = 18. By observing the figures some preliminary conclusions can be inferred:

- The test statistic yields two different clouds of points, then it is able to discriminate between nominal and faulty conditions, when the two clouds are well separated;
- The cloud separation depends mainly on b;
- The cloud position depends on the geometry. In fact a bias in the satellite with ID = 18 yields a radial error much lower than the one induced by satellite with ID = 7.

Similar diagrams can be done also for a single position coordinate (in both ECEF and ENU), for the horizontal radial error, and for the vertical error. Vertical and horizontal errors are generally considered in aviation applications.

4.1.2.2 The Concept of Slope

We observe now that the ratio between the deterministic radial error $||\hat{\boldsymbol{\delta}}\mathbf{x}_R||^2$ and the deterministic test statistic $r_{LS,b,i}$, both due to the bias, given by

$$S_{\text{slope},i} = \frac{||\hat{\boldsymbol{\delta}}\mathbf{x}_R||^2}{r_{LS,b,i}} = \frac{\sqrt{(A_{1i}^2 + A_{2i}^2 + A_{3i}^2)(N_{sat} - 4)}}{\sqrt{1 - P_{ii}}} \tag{4.10}$$

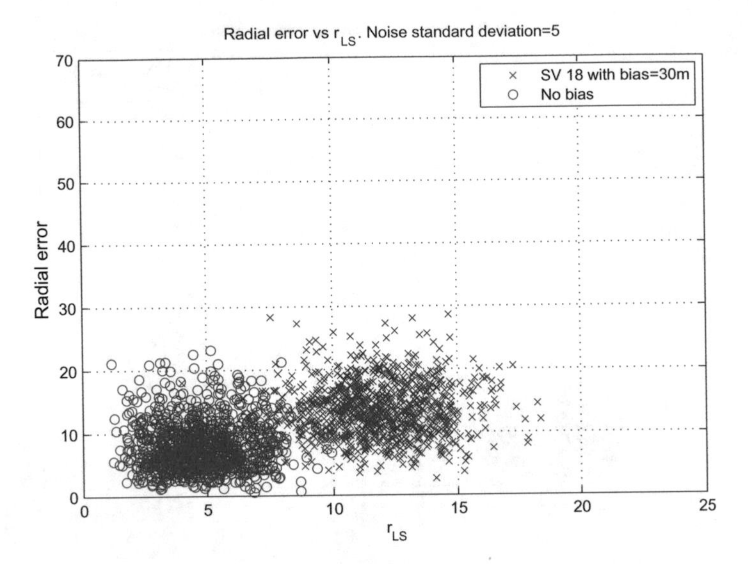

Fig. 4.5 Radial error versus test statistic. Comparison between nominal condition (marker o) and bias in the satellite with ID = 18 (marker x). The bias is 30 m

does not depend on b, but only on the geometry. In literature this ratio is indicated as *slope*, since it identifies a straight line of the type

$$||\hat{\boldsymbol{\delta}}\mathbf{x}_R||^2 = S_{\text{slope},i} r_{LS,b,i}$$

which relates the deterministic components of the radial error and the test statistic, and the slope of the straight line is the ratio $S_{\text{slope},i}$ given in (4.10).

If now we consider both the random and the deterministic components of the error the relationship between the radial error and the test statistic r_{LS} becomes a cloud of points as just seen in the figures of Sect. 4.1.2.1. The presence of the deterministic component is indicated in the Figs. 4.6 and 4.7 with a squared marker obtained for a fixed value of the bias, in particular $b = 50$ m. It is interesting to notice that the same value of bias yields to different situations, depending of the satellite ID. In the configuration with ID = 7, the radial error in the presence of bias is much larger that in the configuration with ID = 18. Moreover if we set the threshold of the test statistic $T_{th,p} = 10$ we observe few events of false alarm in the case of no bias, while the number of events of miss detection (corresponding to the event $r_{Ls} < T_{th,p} = 10$, when bias is present) is greater with ID = 7, than with ID = 18. In few words the configuration with greater slope is worse with ID = 7, both in terms of radial erros and miss detection. Therefore we can claim that the worse configuration is the one with the maximum slope

$$S_{\text{slope},max} = \max_i S_{\text{slope},i}$$

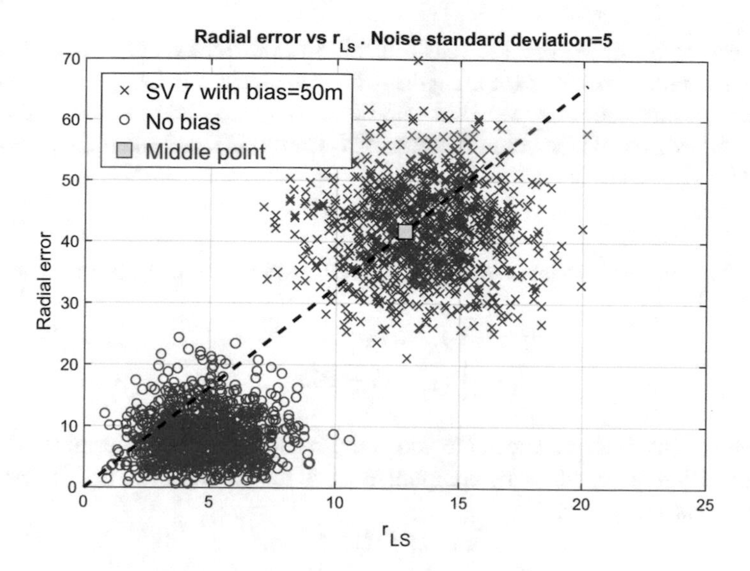

Fig. 4.6 Radial error versus test statistic. Comparison between nominal condition (marker o) and bias in the satellite with ID = 7 (marker x). The bias is 50 m

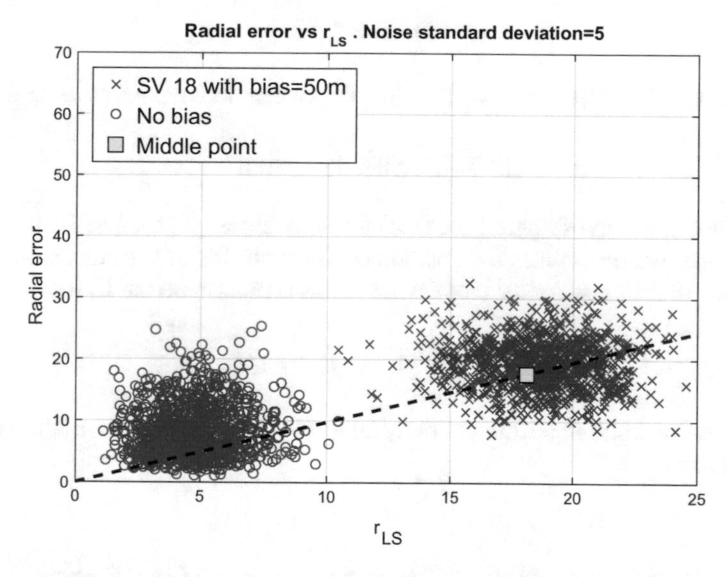

Fig. 4.7 Radial error versus test statistic. Comparison between nominal condition (marker o) and bias in the satellite with ID = 18 (marker x). The bias is 50 m

We can try to determine the reason behind the different behaviour of the test statistic and radial error by exploiting the formalism derived in Chap. 3. We can recall that the measurement error vector $\boldsymbol{\nu}$ can be decomposed into two complementary orthogonal subspaces: the column range of $\mathbf{H}$, $Range\{\mathbf{H}\}$, and the null space of $\mathbf{H}^T$, $Null\{\mathbf{H}^T\}$:

$$\boldsymbol{\nu} = \boldsymbol{\nu}_H + \boldsymbol{\nu}_\perp \tag{4.11}$$

where the two components can be derived using the projection matrix $\mathbf{P} = \mathbf{H}(\mathbf{H}^T\mathbf{H})^{-1}\mathbf{H}^T$

$$\begin{cases} \boldsymbol{\nu}_H = \mathbf{P}\boldsymbol{\nu} \\ \boldsymbol{\nu}_\perp = (\mathbf{I} - \mathbf{P})\boldsymbol{\nu} \end{cases} \tag{4.12}$$

The component $\boldsymbol{\nu}_H$ is undetectable and is the component that determines the position estimation error, while the component $\boldsymbol{\nu}_\perp$ is the residual vector. The estimated position error is equal to

$$\hat{\boldsymbol{\delta}}\mathbf{x} = (\mathbf{H}^T\mathbf{H})^{-1}\mathbf{H}^T\boldsymbol{\nu}_H \tag{4.13}$$

On the other hand the component that we observe with the residual vector is equal to

$$\hat{\boldsymbol{\varepsilon}}_{rr} = (\mathbf{I} - \mathbf{H}(\mathbf{H}^T\mathbf{H})^{-1}\mathbf{H}^T)\boldsymbol{\nu} = \boldsymbol{\nu}_\perp \tag{4.14}$$

Since we know that $||\boldsymbol{\nu}||^2 = ||\boldsymbol{\nu}_\perp||^2 + ||\boldsymbol{\nu}_H||^2$ we can derive the following equality

$$||\mathbf{H}\hat{\boldsymbol{\delta}}\mathbf{x}||^2 + ||\hat{\boldsymbol{\varepsilon}}_{rr}||^2 = ||\boldsymbol{\nu}||^2 \tag{4.15}$$

Under the hypothesis of single fault, the deterministic component $\boldsymbol{b}$ of the error vector $\boldsymbol{\nu}$ is the one that determines the position of the centroids of the clouds. In fact the coordinate of the center of the cloud when a bias is present on the i_{th} pseudorange is

$$(b^2(1 - \mathbf{h}_i\mathbf{G}\mathbf{h}_i^T)), \quad b\sqrt{(\mathbf{h}_i * \mathbf{G}_{1:3}^2 * \mathbf{h}_i^T)} \tag{4.16}$$

The concept of worst slope is thus strictly related to the worst position of the centroids of the clouds.

4.1.3 Confidence Interval Computation in a Slope-Based RAIM

A possible approach to the computation of PL is the so-called slope-based method, which can be described with the support of Fig. 4.8, where four diagrams are shown. In the diagrams (B), (C), and (D) the horizontal axes represent the value of the test statistic used in the FDE test, which is generally $r_p = r_{LS}$. The diagram (B)

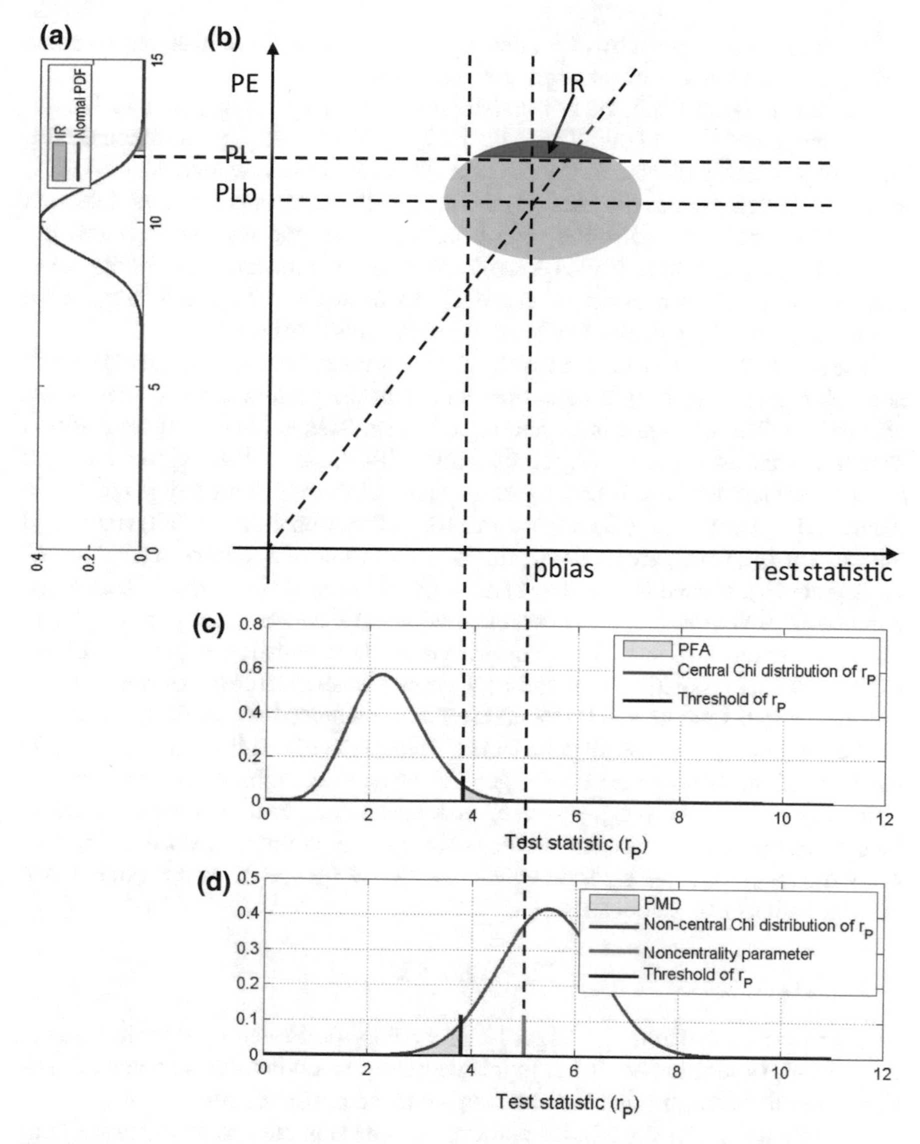

Fig. 4.8 Protection level computation in a classic RAIM scheme

represents the relationship between the test statistic and the absolute position error (PE) in a single coordinate, in the case of maximum slope (the straight line is related to the deterministic components and represents the most dangerous measurement to be faulty). The diagram (C) represents the central chi distribution of the test statistic, when only AWGN is present in the measured pseudoranges. The diagram (D) represents the non-central chi distribution of the test statistic, when both AWGN and a single bias are present in the measured pseudoranges. The non-centrality parameter

$\lambda_{r,i}$ in diagram (D), given by (4.9), depends on the value of the bias, on the noise variance, on the number of satellites and on the geometry.

As shown in Sect. 3.2.3, the test statistic is compared against a threshold $T_{th,P}$ to detect the presence of a fault. The value of $T_{th,P}$ is set by assigning a target false alarm probability, indicated in the diagram (C) with the yellow area denoted PFA, while the yellow area in diagram (D) denoted PMD represents the miss detection probability. The "miss detection" event happens when the test statistic does not exceed the threshold, but a fault is present with a contribution too small for the test to pass. The value of the miss detection probability depends on $T_{th,P}$ and on the value of the non-centrality parameter of the non-central chi-distribution.

The role of diagram (B) is to convert the values related to the test statistic (which is an observable quantity) to equivalent values of the position error (which is not observable). The final goal is to determine the confidence interval of the position error associated to a specific target probability of IR (the so-called PL). At this point the first problem we have is how to set the value of the non-centrality parameter in diagram (D). The idea is to fix a target miss detection probability (PMD) and to find the value of $\lambda_{r,i}$, which gives this target PMD. This value of $\lambda_{r,i}$, indicated as "pbias" in diagram (B), is used to find the absolute position error PLb, due to a bias in the absence of random errors. If we assume that, at the most, one faulty measurement remains after the FDE test, PLb is considered the deterministic component with the most impact in the position error. Notice that pbias is not the true (unknown) value of the non-centrality parameter. It is a value able to guarantee to reach the target PMD.

We are now ready to use diagram (A) to evaluate the PL. This diagram represents the PDF of the absolute error of a generic position coordinate x, which can be written as $|\varepsilon_{x,b}| = |b_x + \varepsilon_x|$, where b_x is a bias and ε_x is a zero mean Gaussian random component. If the standard deviation σ_x of ε_x is much lower than $|b_x|$, we can write $|\varepsilon_{x,b}| \cong |b_x| + \varepsilon_x$, from which $|\varepsilon_{x,b}| \sim \mathcal{N}(|b_x|, \sigma_x)$. At this point, when the bias $|b_x|$ is PLb, we obtain

$$\mathrm{PL} = \mathrm{PLb} + k\sigma_x$$

where k is obtained from the target IR probability (indicated with a blue area in diagram A), as seen in Sect. 3.1.1. In this way the protection level is computed such as to cover the situation with the most impact in the position error.

The rational of the slope-based approach can be summarized as follows. Let us consider a situation where the pseudorange bias is in the satellite with the maximum slope and the corresponding test statistic has a non-centrality parameter equal to pbias. The probability that the RAIM algorithm does not detect this bias is of course PMD. Larger biases lead to a higher non-centrality parameter, shifting the PDF on the right, so they are always detected with a miss detection probability lower than PMD. In this sense pbias is the maximum "undetected" (or the minimum "detected") non-centrality parameter, that is the non-centrality parameter which leads to the maximum undetected (or the minimum detected) position error. On the contrary smaller bias magnitudes have smaller non-centrality parameters, corresponding to a probability of not detection greater than PMD. Fortunately, they also lead to lower positioning errors that compensate the higher non-detection probability. Therefore the PL evaluated from pbias is considered a conservative estimate of the confidence interval, satisfying the requirements on target IR probability, false alarm and miss detection probabilities. Similar consideration can be found also in [3].

The slope-based technique just described can be generalized to a situation when up to N measurements can be faulty at the same time. The only difference is the method used to compute the maximum slope. Instead of searching for the satellite with the maximum slope, one has to search for a linear combination of satellites that yields the maximum slope. With small modifications the method can be also used for radial position errors, such as the horizontal error.

4.1.4 Fault Detection Algorithms in a Slope-Based RAIM

The slope concept can be also used in the design of the fault detection algorithms.

In [3] the authors propose a method which consists of two modules, i.e. the RAIM availability check and the FD. In the first module the slope of each pseudorange measurement is calculated at each epoch. Then the value of PLb is computed by using the maximum slope and the corresponding chi-squared noncentrality parameter. If the PLb exceeds the AL, RAIM stops working because it cannot monitor integrity with the required AL, i.e., with the assigned PMD and PFA. As a consequence the estimated position is declared not valid. If the PLb is equal to or lower than the AL, the RAIM can work and proceeds to check whether the estimated position is faulty or not. The test statistic is computed, and compared against the corresponding detection threshold. If a fault is detected, the position cannot be used. When the test is lower than the threshold, the position is declared valid.

4.2 Method to Evaluate an Upper Bound for PL

When the nominal conditions are not verified, the literature on aviation suggests to compute a single PL for each faulty case and then select the maximum as the final value. The following sections give more details on this methodology, and explain the reasoning behind the approach used in aviation systems. What is more interesting is that this reasoning can be used as a starting point to define other methods of PL computation especially tailored for land applications. In fact it clarifies the theoretical background of the method, why the adopted approximation leads to a conservative solution, ad why many papers are present in the literature devoted to the reduction of conservatism on the PL computation.

The method consists in the evaluation of an upper bound for PL. We start in Sect. 4.2.1 with an example with two possible hypotheses (fault-free and faulty conditions), and we will generalize the method to the cases of multiple faults in Sect. 4.2.3.

The description of the method refers to the computation of the VPL for the vertical component of the estimated position, but the method can be also applied to any single position coordinate. Similar methods are used for the computation of the HPL, which can be found in the literature, for example in [4].

4.2.1 Nominal and Faulty Conditions: Binary Hypothesis Case

We start by considering an example with two possible cases (or hypotheses):

$$\mathscr{H}_0 : \text{Nominal condition}$$
$$\mathscr{H}_a : \text{Fault condition (not detected)}$$

characterized by the corresponding probabilities that these events occur:

$$\Pr(\mathscr{H}_0) : \text{Nominal condition probability}$$
$$\Pr(\mathscr{H}_a) : \text{Not detected fault probability}$$

The event $\mathscr{H}_a$ occurs when a fault is present and it is not detected, thus its probability is:

$$\Pr(\mathscr{H}_a) = P_{\text{MD},1} P_{\text{prior},1} \tag{4.17}$$

where:

- $P_{\text{MD},1}$ is the probability of not detecting a fault, when this occurs (i.e. such a probability is a probability of missed detection, conditioned by the fault occurrence);
- $P_{\text{prior},1}$ is the prior probability [5] to have a fault (sometimes called *fault probability*).

Generally, $\Pr(\mathscr{H}_a)$ is small (e.g. 10^{-5}), therefore $\Pr(\mathscr{H}_0) = 1 - \Pr(\mathscr{H}_a)$ is approximately equal to 1.

4.2.2 Hypothesis $\mathscr{H}_a$: Error Model Based on Non Zero Mean Gaussian PDF

This section describes how to compute the VPL, when the random variable E_m, representing the error of the vertical position coordinate at the mth epoch, can be modelled as:

- a Gaussian random variable, with zero mean and variance $\sigma^2_{m,0}$, in the $\mathscr{H}_0$ hypothesis;
- a Gaussian random variable, with mean $\mu_{m,1}$ not null and variance $\sigma^2_{m,1}$, in the $\mathscr{H}_a$ hypothesis.

The definition of the model implies that we know (or we can estimate) the values of both $\sigma^2_{m,0}$ and $\mu_{m,1}$. In practice this is not an easy task, and represents a critical point. We will address this issue in Sects. 4.2.4 and 4.2.5. For the time being we ignore this fact and we imagine to know these values.

Following the same approach used in Sect. 3.1.1, it is possible to compute:

$$P_m(T_m) = \Pr(|E_m| > T_m) \tag{4.18}$$

where $|E_m| > T_m$ represents the event of IR [5], or risk event, and T_m is a (positive) threshold value. Notice that in Sect. 3.1.1 the used symbol was P_T, here instead of writing $P_{T,m}$ we lightened the notation and decided to call the variable P_m, and the same applies to T_m that was called T_e. The key point here is the computation of this threshold, taking into account the possibility to have both conditions (nominal and faulty). The first step to compute the threshold is to rewrite (4.18) in the form

$$P_m(T_m) = \Pr(|E_m| > T_m|\mathscr{H}_0)\Pr(\mathscr{H}_0) + \Pr(|E_m| > T_m|\mathscr{H}_a)\Pr(\mathscr{H}_a) \tag{4.19}$$

If the fault induces a bias $\mu_{m,1}$, $P_m(T_m)$ becomes:

$$\begin{aligned} P_m(T_m) = {} & \Pr(\mathscr{H}_0)\operatorname{erfc}\left(\frac{T_m}{\sqrt{2\sigma^2_{m,0}}}\right) + \\ & \frac{\Pr(\mathscr{H}_a)}{2}\left\{\operatorname{erfc}\left(\frac{T_m - |\mu_{m,1}|}{\sqrt{2\sigma^2_{m,1}}}\right) + \operatorname{erfc}\left(\frac{T_m + |\mu_{m,1}|}{\sqrt{2\sigma^2_{m,1}}}\right)\right\} \end{aligned} \tag{4.20}$$

where the absolute value of $\mu_{m,1}$ is introduced to highlight that the obtained expression of $P_m(T_m)$ is an even function of $\mu_{m,1}$. It is important to stress that (4.20) is valid only in the hypothesis of a Gaussian model with mean different from zero.

4.2.2.1 VPL Computation from Eq. (4.20)

In order to compute a valid VPL, by taking into account the two hypothesis $\mathscr{H}_0$ and $\mathscr{H}_a$, we have to start from an initial IR probability $P_{m,\text{target}}$ (as obtained from the IR requirements) and invert (4.20) to compute the value of $T_{m,\text{bound}}$ that satisfies the relationship $P_m(T_{m,\text{bound}}) = P_{m,\text{target}}$. However, it is evident that the inversion of (4.20) is not straightforward. In Sect. 3.1.1 we have seen that the evaluation of the threshold $T_{m,\text{bound}}$ in the case of nominal conditions leads to the formula:

$$T_{m,\text{bound}} = \sqrt{2}\operatorname{erfcinv}(P_{m,\text{target}})\sigma_{m,0} = k(P_{m,\text{target}})\sigma_{m,0} \tag{4.21}$$

where the key factor $k(P_{m,\text{target}})$ only depends on the desired target probability. However this simple expression cannot be used to invert (4.20). A method commonly used in aviation to provide an expression for the inversion of (4.20), which can be easily computed at each epoch, is based on the idea to divide the $P_{m,\text{target}}$ in two parts:

$$
\begin{aligned}
P_{m,\mathscr{H}_0,\text{target}} &= \alpha P_{m,\text{target}} \\
P_{m,\mathscr{H}_a,\text{target}} &= (1-\alpha) P_{m,\text{target}}
\end{aligned}
$$

and to associate these target probabilities to the events $\mathscr{H}_0$ and $\mathscr{H}_a$. Such probabilities are at the basis for the definition of the so-called *fault tree* that will be described in details in Sect. 4.2.6, considering a more general case of multiple hypotheses.

Again the inversion of the part of (4.20) related to $\mathscr{H}_a$ is not trivial. A simpler alternative is to invert all the three terms of (4.20) with the probability partitioning

$$
P_{m,\mathscr{H}_0,\text{target}} \quad 0.5 P_{m,\mathscr{H}_a,\text{target}} \quad 0.5 P_{m,\mathscr{H}_a,\text{target}}
$$

by obtaining

$$
T_m(\mathscr{H}_0) = \sigma_{m,0}\sqrt{2}\text{erfcinv}\left(\frac{P_{m,\mathscr{H}_0,\text{target}}}{\Pr(\mathscr{H}_0)}\right) \tag{4.22}
$$

$$
T_m(\mathscr{H}_{a+}) = |\mu_{m,1}| + \sigma_{m,1}\sqrt{2}\text{erfcinv}\left(\frac{P_{m,\mathscr{H}_a,\text{target}}}{\Pr(\mathscr{H}_a)}\right) \tag{4.23}
$$

and

$$
T_m(\mathscr{H}_{a-}) = |\mu_{m,1}| - \sigma_{m,1}\sqrt{2}\text{erfcinv}\left(\frac{P_{m,\mathscr{H}_a,\text{target}}}{\Pr(\mathscr{H}_a)}\right) \tag{4.24}
$$

that are valid under the assumption of Gaussian models.

These partial $T_m(\cdot)$ are different from the term $T_{m,\text{bound}}$ that we would obtain if we directly inverted (4.20). A proposed method, available in the scientific literature, is to set as bound the following parameter:

$$
T_{m,\text{bound}}^{\max} = \max\{T_m(\mathscr{H}_0), T_m(\mathscr{H}_{a+})\} \tag{4.25}
$$

by ignoring the term $T_m(\mathscr{H}_{a-})$, since $T_m(\mathscr{H}_{a+})$ is always greater than $T_m(\mathscr{H}_{a-})$. This is a conservative choice with respect to the direct inversion of (4.20). In fact, since we know that the *erfc* is a decreasing function (i.e. it represents the area of the tail of the Gaussian PDF), we can write the following expression:

$$
P_m(T_{m,\text{bound}}^{\max}) < P_{m,\mathscr{H}_0,\text{target}} + P_{m,\mathscr{H}_a,\text{target}} = P_{m,\text{target}} \tag{4.26}
$$

This equation means that if we select the quantity $T_{m,\text{bound}}^{\max}$ as bound of the confidence interval, the IR probability as stated by $P_m(T_{m,\text{bound}}^{\max})$ is smaller than the target probability $P_{m,\text{target}}$.

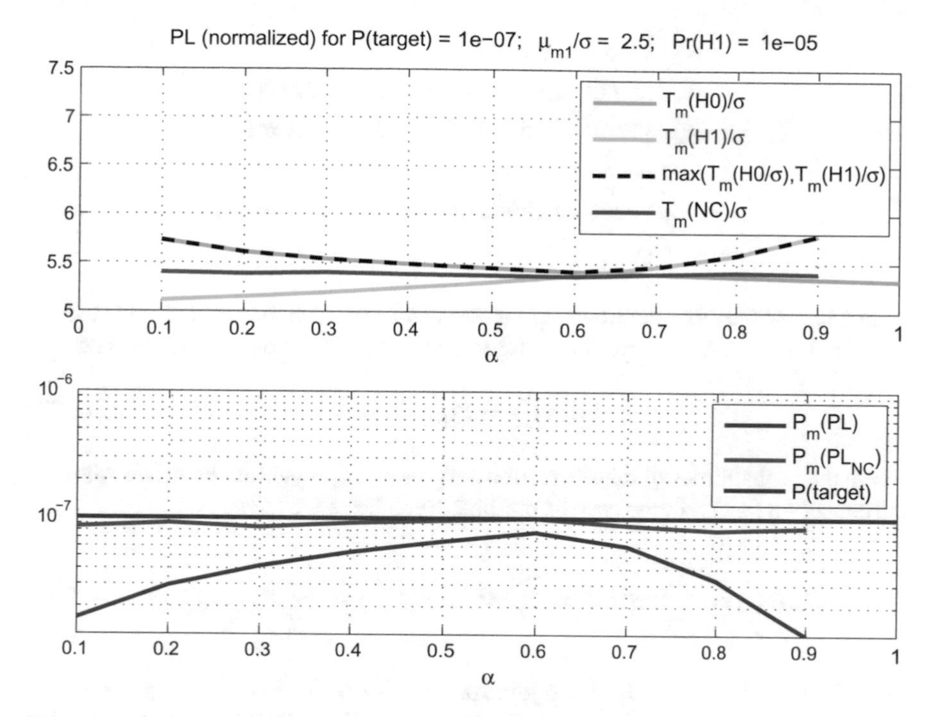

Fig. 4.9 Conservatism analysis

The level of conservatism introduced by the method (4.25) can be investigated by simulation. Figure 4.9 shows the values of $T_m(\mathscr{H}_a)$ and $T_m(\mathscr{H}_{0+})$ normalized with respect to the variance (assumed equal in the two cases) versus α. The other simulation parameters are $\Pr(\mathscr{H}_a) = 10^{-5}$, $\mu_{m,1}/\sigma_{m,1} = 2.5$, and $P_{m,\text{target}} = 10^{-7}$. In the figure the value of T_m obtained by numerical inversion of (4.20) is compared with the value of $T^{\max}_{m,\text{bound}}$, showing a moderate level of conservatism. Similar results can be found with different values of the parameters. The graph at the bottom of the figure shows the probability obtained with the VPL obtained by numerical inversion of (4.20) compared with the one associated to $T^{\max}_{m,\text{bound}}$. It is evident that the first one coincides with the target probability, while the second one is lower.

4.2.3 *Confidence Interval in the Case of Multiple Faults*

The scientific literature considers multiple hypothesis in order to compute valid VPL measurements:

$\mathscr{H}_0$: Nominal condition

$\mathscr{H}_{a1}$: Fault condition of type 1 (not detected)

$\mathscr{H}_{a2}$: Fault condition of type 2 (not detected)

$\cdots : \cdots$

$\mathscr{H}_{aj}$: Fault condition of type j (not detected)

$\cdots : \cdots$

For each of the aforementioned hypothesis we have to set the probability $\Pr(\mathscr{H}_{aj})$, for $j = 0, 1, \cdots, N_f$, where N_f represents all the possible types of faults, and:

$$\Pr(\mathscr{H}_{aj}) = P_{\text{MD},j} P_{\text{prior},j} \tag{4.27}$$

where the probabilities of miss detection and the prior probability have been already introduced in (4.17). In the case of multiple hypothesis we have:

$$P_m(T_m) = \sum_{j=0}^{N_f} \Pr(|E_m| > T_m | \mathscr{H}_{aj}) \Pr(\mathscr{H}_{aj}) = \sum_{j=0}^{N_f} P_{m,j}(T_m) \tag{4.28}$$

This Equation is valid only if the joint probability of having more than one fault not detected is equal to zero (single failure assumption). Practically speaking, this probability is considered negligible and (4.28) is typically used to compute $P_m(T_m)$. Equation (4.28) is called *rare event approximation* and is accurate to within about 10% of the true probability when $P_{m,j}(T_m) < 0.1$. The method that is used to calculate the $T_{m,\text{bound}}$ (i.e.: basically the VPL) is based on the initial setting of the following parameters:

- $P_{m,\text{target}}$;
- $P_{m,j,\text{target}}$ (i.e.: every single contribution given by $\mathscr{H}_{aj}$ to the $P_{m,\text{target}}$);
- $\Pr(\mathscr{H}_{aj})$.

This operation, which is also called *integrity allocation*, is performed through the Fault Tree method. In conclusion we can say that, for every contribution of (4.27), we compute the parameter $T_m(\mathscr{H}_{aj})$ and we adopt as bound of the confidence interval related to the $P_{m,\text{target}}$ (that is, the VPL) the expression

$$T_{m,\text{bound}}^{\max} = \max\{T_m(\mathscr{H}_0), T_m(\mathscr{H}_{a1}), \cdots, T_m(\mathscr{H}_{aj}), \cdots, T_m(\mathscr{H}_{N_f})\} \tag{4.29}$$

When the error conditioned by different hypotheses $\mathscr{H}_{aj}$ (with $j \neq 0$) can be modelled as a Gaussian random variable with mean μ_m and variance $\sigma_{m,j}^2$, the parameters $T_m(\mathscr{H}_0)$ and $T_m(\mathscr{H}_{aj})$ can be computed by inverting the erfc function, similarly to the case shown in Sect. 4.2.1 for the binary problem. If we keep the same approach used to derive (4.23) we will have:

$$T_m(\mathscr{H}_{aj}) = |\mu_{m,j}| + \sigma_{m,j}\sqrt{2}\text{erfcinv}\left(\frac{P_{m,j,\text{target}}}{\Pr(\mathscr{H}_{aj})}\right) \tag{4.30}$$

Eventually, we can say that $T_{m,\text{bound}}^{\max}$ is conservative with respect to $T_{m,\text{bound}}$. Methods aimed at reducing such conservative approach are available in the scientific literature.They are generally tailored for a specific applications, and cannot be generalized.

Notice that the results derived in this section cannot be applied if the rare event approximation is not valid, or the error associated to a generic hypothesis $\mathscr{H}_{aj}$ is not Gaussian, or cannot be modeled. In these cases ad hoc methods have to be envisaged, generally valid only for the specific application scenario under study.

4.2.4 Estimation of the Mean of the Gaussian Error Model

The value of $\mu_{m,j}$ in (4.30) are generally not known and have to be estimated in some way. We describe here a possible method valid when the multiple hypotheses are:

$$
\begin{aligned}
&\mathscr{H}_0 : \text{Nominal condition (Fault free)}\\
&\mathscr{H}_{a1} : \text{Single fault in satellite 1 (not detected)}\\
&\mathscr{H}_{a2} : \text{Single fault in satellite 2 (not detected)}\\
&\ldots : \ldots\\
&\mathscr{H}_{aj} : \text{Single fault in satellite j (not detected)}\\
&\ldots : \ldots
\end{aligned}
$$

This means that we consider a case where at each epoch m only a satellite is faulty. Moreover we add the hypothesis that the fault induces a bias $\mu_{m,j}$ in the position coordinate we are analysing, while the random noise is white and Gaussian.

If x is the generic coordinate we are considering, the error E_m at the m-th epoch can be written as

$$E_m = \hat{x}_0 - x_t$$

where $\hat{x}_0$ is the position estimate obtained with all the satellites in view, and x_t is the true position coordinate, and where the dependence on time instant m of these two quantities has been dropped for notational simplicity. By adopting the same approach used in the solution separation test, we evaluate for each hypothesis $\mathscr{H}_{aj}$ the position coordinate $\hat{x}_j$ obtained by excluding the satellite j. The event $\{|E_m| > T_m|\mathscr{H}_{aj}\}$ involved in (4.28) can be written as

$$\{(|E_m| > T_m)|\mathscr{H}_{aj}\} = \{(|\hat{x}_0 - x_t| > T_m)|\mathscr{H}_{aj}\} = \{(|\hat{x}_j + \hat{\mu}_{m,j} - x_t| > T_m)|\mathscr{H}_{aj}\}$$

where

$$\hat{\mu}_{m,j} = \hat{x}_0 - \hat{x}_j \tag{4.31}$$

It is evident that if we consider only the deterministic components of the estimated coordinates (that is the values when the random noise is zero), $\hat{\mu}_{m,j} = \mu_{m,j}$, while in general $\hat{\mu}_{m,j}$ is an estimate of $\mu_{m,j}$, as given in (3.27). At this point we can write the probabilities in (4.28) in the form

$$\Pr(|E_m| > T_m|\mathscr{H}_{aj}) = \Pr(|\hat{x}_j + \hat{\mu}_{m,j} - x_t| > T_m|\mathscr{H}_{aj})$$

Since

$$|\hat{x}_j + \hat{\mu}_{m,j} - x_t| \leq |\hat{x}_j - x_t| + |\hat{\mu}_{m,j}|$$

we can write

$$\Pr(|\hat{x}_j + \hat{\mu}_{m,j} - x_t| > T_m|\mathscr{H}_{aj}) \leq \Pr(|\hat{x}_j - x_t| + |\hat{\mu}_{m,j}| > T_m|\mathscr{H}_{aj})$$

To evaluate $T_m(\mathscr{H}_{aj})$ we write

$$\Pr(|\hat{x}_j + \hat{\mu}_{m,j} - x_t| > T_m|\mathscr{H}_{aj}) \leq \Pr(|\hat{x}_j - x_t| + |\hat{\mu}_{m,j}| > T_m|\mathscr{H}_{aj}) = \frac{P_{m,j,\text{target}}}{\Pr(\mathscr{H}_{aj})}$$

from which, by adopting the same approach used to derive (4.23), and by using (4.31), we find that a conservative solution for the evaluation of $T_m(\mathscr{H}_{aj})$ is

$$\begin{aligned} T_m(\mathscr{H}_{aj}) &= |\hat{\mu}_{m,j}| + \sigma_{m,j}\sqrt{2}\text{erfcinv}\left(\frac{P_{m,j,\text{target}}}{\Pr(\mathscr{H}_{aj})}\right) \\ &= |\hat{x}_0 - \hat{x}_j| + \sigma_{m,j}\sqrt{2}\text{erfcinv}\left(\frac{P_{m,j,\text{target}}}{\Pr(\mathscr{H}_{aj})}\right) \end{aligned} \tag{4.32}$$

This value can be used in (4.29) to evaluate the VPL. A comparison between this method based on the solution separation test and a method based on the slope concept can be found in [6].

Notice that the result in (4.32) is valid for the case of a single failure, but the method can be generalized to the case of simultaneous multiple failures.

4.2.5 *Modelling of the Variance of the Gaussian Error Model*

Modelling the variance of the Gaussian error model can be a complex task, and models can vary hugely depending on the considered application. However, at the moment, the model derived for avionic application [7] is used also in different applications such as the maritime ones [4].

We describe here only the method used in aviation, which is based on the availability of additional data provided by the so-called Integrity Support Message (ISM), [4]. This is a low rate message sent by ground stations to the receivers with information about the maximum bias for each measured pseudorange, and the variance of the satellite clock and ephemeris errors. We do not give here the details of this message and on how the ISM communication is implemented, because it is out of the scope of this book. However, it is interesting to know that the problem of assigning a value to the variance has been solved in aviation by using the support of a ground station, and this aspect has to be considered when integrity operations have to be implemented in other applications.

In aviation applications the variance of the i_{th} pseudorange (the i_{th} element of the diagonal of the matrix $\mathbf{\Sigma}$) is computed by a double frequency receiver as the sum of three components:

$$\sigma^2 = \sigma^2_{URA_i} + \sigma^2_{tropo,i} + \sigma^2_{user,i} \quad [\mathrm{m}^2] \tag{4.33}$$

where

- $\sigma^2_{URA_i}$ is the variance of the clock and ephemeris error of the i-th satellite, provided by ISM.
- $\sigma^2_{tropo,i}$ is the variance of the error due to tropospheric propagation and is computed as:

$$\sigma_{tropo,i} = 0.12 + \frac{1.001}{\sqrt{0.002001 + \sin(\frac{\theta\pi}{180^\circ})^2}} \quad [\mathrm{m}] \tag{4.34}$$

  where θ is the elevation angle of the satellite in degrees.
- $\sigma^2_{user,i}$ is the user component of the total variance. It is a composition of the thermal noise and the multipath induced noise, expressed as

$$\sigma_{user,i} = \frac{f^4_{L1} + f^4_{L5}}{(f^2_{L1} - f^2_{L5})^2}\sqrt{\sigma^2_{MP} + \sigma^2_{noise}} \quad [\mathrm{m}^2] \tag{4.35}$$

  where f_{L1} and f_{L5} are the carrier frequencies of the L1 and L5 bands, and the multipath and noise components are computed respectively as

$$\sigma_{MP} = 0.13 + 0.53\exp\left(-\frac{\theta}{10^\circ}\right) \quad [\mathrm{m}] \tag{4.36}$$

  and

$$\sigma_{noise} = 0.15 + 0.43\exp\left(-\frac{\theta}{6.9^\circ}\right) \quad [\mathrm{m}] \tag{4.37}$$

Notice that the results reported in this section are valid for the user receivers assumed in [4]. For other applications these results have to be checked and possibly modified. Moreover, if the position is computed with the support of an augmentation system, other contributions to the variance have to included, as described in Chap. 5.

4.2.6 Fault Tree

It is not straightforward to summarize in a simple section the methodology known as Fault Tree Analysis (FTA). For this reason, this section is focused on some of the main aspects related to the FTA introduced in [5] for the Local Area Augmentation System (LAAS) and oriented to the PL computation in the case of multi-hypotheses.

FTA is a technique for the failure analysis which focuses on one specified undesired event (in our case the event of loss of integrity) and which provides a method for determining conditions and factors that can cause the failure. The undesired event represents the top event in a fault tree diagram, where the contributors to the undesired event are identified and organized in a logical manner and represented pictorially. In particular the fault tree shows the inter-relationships of the basic events that lead to the undesired event. Notice that the identified faults are not generally exhaustive, as they cover only the most credible faults as assessed by the analyst. At this point it is better to make a distinction between the two words failure and fault:

- a failure is a malfunction of a device, due to something wrong generated inside the device;
- a fault is a malfunction of a device due to an external wrong input.

Therefore all the failures generate faults, but not all faults are due to device failures.

A typical fault tree is composed by a number of elements, which can be grouped as follows:

- Elements which represent events (basic, intermediate and top event);
- Elements which represent "gates" (they indicate the relationships among events). The most important gates are:
  - OR-GATE: output fault occurs if at least one of the input faults occurs
  - AND-GATE: output fault occurs if all of the input faults occur

A simple fault tree is shown in Fig. 4.10, where the events are indicated just as Q, A, B, C, and D. In more complex cases other symbols could be present in a fault tree. A complete list of such symbols can be found in [8].

In summary the main steps to draw a fault tree are:

- Identify the top event;
- Identify the intermediate and initiating events;
- Assign the probabilities to the initiating events.

Once the fault tree is drawn, the computation of the probability of occurrence of the top event (and of any internal event corresponding to a logical sub-system) can be performed on the basis of the probabilities assigned to the basic (or initiating) events, which represent failure events of the basic components. In the approach described in Sect. 4.2.3, the top event is the loss of integrity (LOI), the hypotheses (which contribute to the top event "LOI") have been assumed mutually exclusive (the event of multiple faults is not considered), then only OR gates are present in the FTA. For this reason in these cases some authors omit the OR-GATEs in the pictorial representation of the fault tree (2.11).

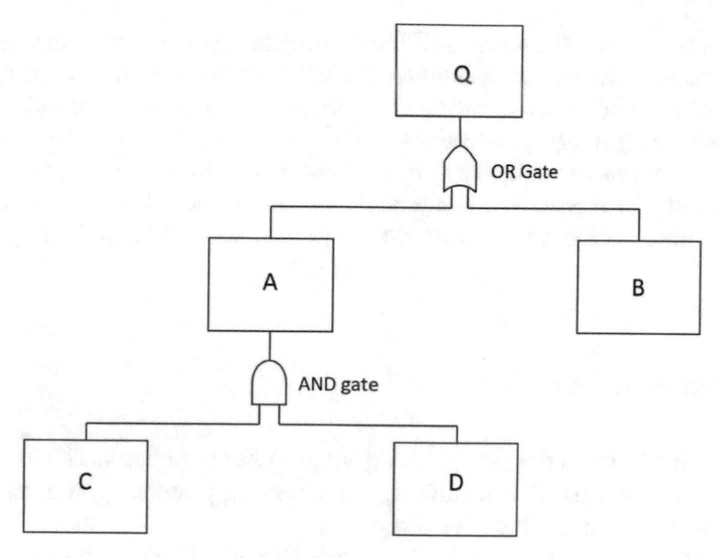

Fig. 4.10 Example of fault tree

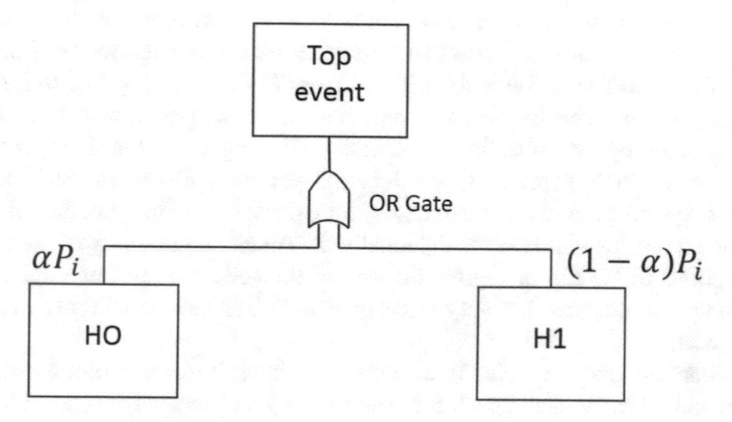

Fig. 4.11 A simple fault tree

4.2.6.1 IR Allocation and PL Computation

The method for the PL evaluation, presented in Sect. 4.2.1, is based on the idea of the sub-allocation to the two cases $\mathscr{H}_0$ and $\mathscr{H}_a$, computing different PLs. This situation can be represented, by the fault tree of Fig. 4.11, where the top event is the LOI.

The concept of "sub-allocation of the IR" is present in many papers related to GBAS (and more specifically to LAAS) and it is the way normally used for the PL evaluation, especially in the case of multi-hypotheses tree. As written in [9], the process of creating distinct PLs for each event of the fault tree simplifies integrity analysis. The combined error distribution, covering both nominal and faulted conditions, is generally non-Gaussian and difficult to model. On the contrary, decomposing integrity analysis for distinct and nominal conditions makes Gaussian assumptions more reasonable and permits more rigorous bounding of uncertain model parameters.

As stated in [9], integrity allocations could conceptually be allocated in a dynamic, time-varying manner to optimize system performance; however PL values are generally remarkably insensitive to large variations in IR allocations because they are derived from Gaussian error models. Consequently, static allocations are generally used.

4.3 Stanford Plot

The so-called Stanford diagram (or Stanford plot) has been introduced in [10] and presents a valuable tool for monitoring and assessing positioning systems performance (in terms of availability and integrity).

The layout of a generic Stanford diagram is shown in Fig. 4.12. For each sample position and protection level, a point is plotted in the Stanford diagram whose abscissa represents the absolute position error and whose ordinate represents its associated protection level. Usually, separate Stanford plots are represented for the Horizontal Position Error (HPE) and Vertical Position Error (VPE), corresponding to HPLs and VPLs, respectively. The diagonal axis separates those samples in which the position error is covered by the protection level, above the diagonal, from those, below the diagonal, in which the protection level fails to cover the position error. Stanford plots allow an easy and quick check that integrity holds, just by making sure that all sample points lie on the upper side of the diagonal axis. Also, the proximity of the cloud of sample points to the diagonal gives an idea of the achieved level of safety, as any point above the diagonal but very close to it indicates that an integrity event was close to occur.

The Stanford diagram actually accounts for integrity events (not for integrity failures) and allows to distinguish between two types of integrity events [11]:

- Misleading Information (MI) events, and
- Hazardous Misleading Information (HMI) events.

A MI event occurs when, being the system declared available, the position error exceeds the protection level but not the AL. A HMI event occurs when, being the system declared available, the position error exceeds the AL.

The concept of the Stanford diagram has been further improved in [12], introducing the Stanford-ESA integrity diagram. In detail, the Stanford-ESA diagram also includes a 2D histogram showing the relationship of position errors against protection levels for a set of measurements using an all in view satellite selection. Its computation is based on the following steps, performed at a given location and for each instant of time:

- Analyse all the possible satellite combinations, from 4 to all-in-view GPS satellites;

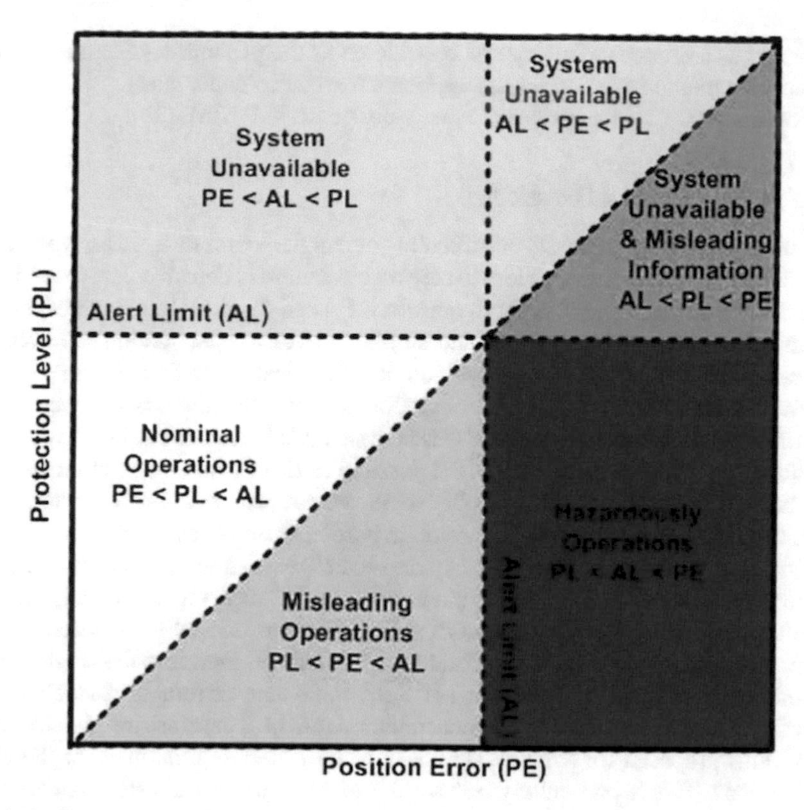

Fig. 4.12 Stanford plot

- For each combination of satellites, compute the HPE, HPL pair and the VPE, VPL pair;
- Include in a Stanford Diagram the (HPE, HPL) pair or the (VPE, VPL) pair, respectively leading to a graph with horizontal or vertical information about the minimum so-called Safety Index.

4.4 Final Remarks on RAIM Algorithms

In this chapter and in Chap. 3 we have seen the operations performed by a RAIM to arrive to a reliable computation of the protection level. In this final section we summarize the main aspects considered so far and we discuss on possible variations of the methods so far described. Notice that the possible variations of the RAIM algorithms are generally scenario-dependent, and at the time of writing this book they are not well-established yet. However, any method aimed to compute the PL cannot ignore the well-established methods used in aviation applications. Therefore the methods so far described can be considered as the preliminary step to arrive to formulate a method specific for the application scenario under study.

To summarize, two questions are generally posed to RAIM, [13]:

1. Does a failure exist?
2. If so, which is the failed satellite?

The answer to question (1) is sufficient for supplemental navigation when there is an alternative navigation system to rely on if a failure is detected. However, in the case of sole-means navigation, both questions (1) and (2) must be answered, so that the failed satellite can be identified and eliminated from the navigation solution, and the navigation can proceed safely with an uncontaminated GNSS solution.

Traditional RAIM, i.e., RAIM for aviation, assumes that there is only one failure at a time (single failure hypothesis). This is reasonable for aviation, since in open sky conditions and with no interference, the possible cause of a range measurement error is a failure of the satellite segment of GNSS, whose probability of occurrence has been kept very low by the system control, in particular in the last years. Nonetheless, in terrestrial applications, the open sky and interference-free hypotheses cannot be guaranteed. For example in urban environments, the receiver is typically in non-Line-Of-Sight (nLOS) or quasi-nLOS with respect to one or more satellites and the presence of an undetected interference is not remote, therefore the single failure hypothesis is no more acceptable [6, 14]. Sometimes authors refuse the single failure hypothesis also in the case of multi-constellation RAIM. Therefore the classical FDE algorithms proposed for aviation [15] have to be generally redefined for the other applications. This is particularly true for the part regarding exclusion. A variety of RAIM techniques have been proposed in the literature, and all are based on some kind of self-consistency check among the available redundant measurements. These techniques fall into three categories:

1. Snapshot RAIM, in which only current redundant measurements are used in the self-consistency check;
2. Snapshot Advanced Receiver Autonomous Integrity Monitoring (ARAIM), that is an evolution of RAIM in which an ISM is received from ground stations;
3. Sequential (or averaging or filtering) RAIM, which uses past as well as present measurements along with assumptions about the receiver motion.

Snapshot schemes have gained more acceptance, as they have a far simpler and better established mathematical treatment. Snapshot RAIM is a natural choice for punctual (e.g., code based) position fixes [16]. On the other hand, sequential schemes are recommended whenever the current measurements (measured pseudoranges) are not independent from the past ones (e.g., in the presence of smoothed pseudoranges, carrier-phase ranging measurements, tight integration with dead-reckoning sensors, e.g., with inertial sensors, or with laser scanner observations, etc..., whose implementations are typically based on Kalman filter). Indeed, in these types of implementations, observations collected within the filtering interval are all vulnerable to rare-event integrity threats, such as user equipment and satellite failures; when trying to protect the system against such faults (which may last in time), snapshot RAIM is limited in that the fault profile over time is never considered [16].

The methods described in this chapter and in Chap. 3, that is

1. the least-squares range residuals [7],
2. the parity method described by Sturza and Brown [17–19],
3. the solution separation test,

are the most commonly known snapshot schemes.

Concerning ARAIM its high-level functional steps can be summarized as follows, [15].

1. Extract the information of the ISM and exclude the satellites set to faulty
2. Estimate the position $\hat{\boldsymbol{\Delta}}\mathbf{x}$ given the set of measured pseudoranges
3. Perform some statistical test to determine the Integrity of the measurements (determine if $\mathscr{H}_0$ hypothesis is accepted)
4. If hypothesis $\mathscr{H}_0$ is accepted go to 5 else
   - if no exclusion is implemented, declare the system as unavailable
   - else, try to perform exclusion. If exclusion is successful return to 3 otherwise declare the system as unavailable
5. Compute the Protection Level and according to the couple (PL,AL) declare the state of the system.

In literature [1] methods for performing exclusion in a sequential way are also described: these methods utilize the information of several successive epochs, in contrast with the single epoch snapshot approach, for performing FDE. Obviously in this case the performance of the algorithm is traded off with complexity.

References

1. Franzese G (2016) GNSS sequential fault detection and exclusion for maritime users. Master's thesis, Politecnico di Torino
2. Zhai Y, Joerger M, Pervan B (2015) Continuity and availability in dual-frequency multi-constellation ARAIM. In: Proceedings of the 28th international technical meeting of the satellite division of the institute of navigation (ION GNSS), Tampa, Florida, p 664–674
3. Salos D, Martineau A, Macabiau C, Bonhoure B, Kubrak D (2014) Receiver autonomous integrity monitoring of GNSS signals for electronic toll collection. IEEE Trans Intell Transp Syst 15(1):94–103
4. Milestone 3 report. Technical report, GPS-Galileo Working Group C ARAIM Technical Subgroup, 26 Feb 2016
5. Pullen S (2001) Tutorial presentation: Augmented GNSS: fundamentals and keys to integrity and continuity. http://www-leland.stanford.edu
6. Blanch J, Walter T, Enge P (2010) RAIM with optimal integrity and continuity allocations under multiple failures. IEEE Trans Aerosp Electron Syst 46(3):1235–1247
7. Parkinson BW, Axelrad P (1988) Autonomous GPS integrity monitoring using the pseudorange residual. NAVIGATION: J Inst Navig (USA), 35(2) (Summer)

8. Vesely WE, Goldberg FF, Haasl DF (1981) Fault tree handbook. Government Printing Office, U.S
9. Gleason S (2009) GNSS applications and methods–GNSS technology and applications. Artech House
10. Stanford GPS lab website. WAAS precision approach metrics: accuracy, integrity, continuity and availability. http://waas.stanford.edu/metrics.html
11. Navipedia. http://www.navipedia.net/index.php/Integrity. 23 Feb 2012
12. Navipedia (2006) The Stanford ESA integrity diagram: focusing on SBAS integrity. http://www.navipedia.net/index.php/The_Stanford_-_ESA_Integrity_Diagram:_Focusing_on_SBAS_Integrity
13. Langley RB (1999) The integrity of GPS. GPS World 10(3):60–63
14. Liu J, Lu M, Cui X, Feng Z (2007) Theoretical analysis of RAIM in the occurrence of simultaneous two satellite faults. IET Radar Sonar Navig 1(2):92–97
15. Blanch J, Walker T, Enge P, Lee Y, Pervan B, Rippl M, Spletter A, Kropp V (2015) Baseline advanced RAIM user algorithm and possible improvements. IEEE Trans Aerosp Electron Syst 51(1):713–732
16. Joerger M, Pervan B (2010) Sequential residual-based RAIM. In: Proceedings of the 23rd international technical meeting of the satellite division of the institute of navigation (ION GNSS 2010), Portland, OR, p 3167–3180
17. Sturza MA, Brown AK (1990) Comparison of fixed and variable threshold RAIM algorithms. In: Proceedings of the 3rd international technical meeting of the satellite division of the institute of navigation (ION GPS 1990), Colorado Spring, CO, p 437–443
18. Sturza MA (1988–1989) Navigation system integrity monitoring using redundant measurements. NAVIGATION: J Inst Navig 35(4):483–502 (Winter)
19. Brown RG (1992) A baseline RAIM scheme and a note on the equivalence of three RAIM methods. NAVIGATION: J Inst Navig 39(3):301–316 (Fall)

Chapter 5
Methods for Protection-Level Evaluation with Augmented Data

Letizia Lo Presti and Marco Pini

Abstract This chapter describes how an augmentation system can support a GNSS receiver of a vehicle (more in general a mobile object) in the PL evaluation. Since this approach has been firstly adopted in aviation, and it is already operative in some airports, we will describe a generic LAAS architecture, as a typical example of a GBAS, tailored to improve the performance of a GNSS receiver in terms of accuracy and integrity. The major components and features of LAAS will be detailed.

5.1 Ground-Based Augmentation Systems

A Ground-Based Augmentation System is a system designed to support ground-based augmentation of the primary GNSS constellations by providing enhanced levels of service for the navigation in safety-critical transport applications. The main goal of GBAS is to increase the accuracy of the user position and to provide integrity data. Nowadays, GBASs are especially used for aviation applications. For this application, the term LAAS is used to indicate a GBAS.

GBAS consists of ground and user equipment. The ground equipment includes generally a number of reference receivers, a ground facility, and a data broadcast transmitter. This ground infrastructure is complemented by GBAS equipment installed on the vehicle. The details of the operations performed by both ground and user equipments are given in the remainder of this chapter.

L. Lo Presti (✉)
Politecnico di Torino, c.so Duca degli Abruzzi n.24, Torino, Italy
e-mail: letizia.lopresti@polito.it

M. Pini (✉)
Istituto Superiore Mario Boella, via P.C. Boggio n.61 Torino, Turin, Italy
e-mail: pini@ismb.it

© Springer International Publishing AG, part of Springer Nature 2018

L. Lo Presti and S. Sabina (eds.), *GNSS for Rail Transportation*, PoliTO Springer Series, https://doi.org/10.1007/978-3-319-79084-8_5

Fig. 5.1 Main component of a local-area augmentation system

5.1.1 LAAS Architecture

Figure 5.1 sketches a LAAS architecture employed to augment the performance of GNSS receivers on board of aircrafts during landing approaches. It is typically composed of:

- static GNSS reference receiver (RRs) with antennas placed at surveyed locations (in Fig. 5.1 they are labeled as GNSS-RR1, GNSS-RR2,. . ., GNSS-RRn);
- the Local Ground Facility (LGF), that gathers measurements from the RRs, computes the pseudorange corrections for each satellite and implements algorithms for integrity monitoring;
- VHF Data Broadcast (VDB) transmitter (in Fig. 5.1 it is labeled as VDB-TX), that sends the pseudorange corrections and integrity data to users on a very-high-frequency (VHF) channel.

The landing aircraft computes its position processing signals received by each satellite vehicle (SV) (in Fig. 5.1 GNSS-SV1, GNSS-SV2,..., GNSS-SVm) and receives augmentation data, PRC, and integrity data, from the LAAS on a dedicated communication channel.

It is evident that the main task of a LAAS is to assist an aircraft during the landing approach. The landing procedure depends on many factors and is rated in thee categories, known as CAT I, CAT II, and CAT III, related to the type of instrument landing system used for the approach, the value of the decision height, and the runway visual range. More details on these categories can be found in [1].

5.2 GNSS Reference Receivers

Generally, RRs are able to perform accurate measurements and are specifically designed for LAAS/GBAS ground station applications. RRs provide code and carrier measurements and embed multipath rejection techniques. They can be single-frequency receivers and could feature specific algorithms for integrity monitoring. An example is the Novatel CMA-4048 [2], which features 24 channels for Signal Quality Monitoring on the GPS L1 C/A code. Eight correlators are available per channel: three of them are fixed and used for tracking purposes, while five correlators are end-user programmable, settable in steps of 25 ns.

RRs antennas are placed in a surveyed position, and in an environment free of multipath and interfering signals. Generally, GNSS antennas used for RRs have a stable phase center and are based on choke ring technology that provides a further barrier against multipath. Today, several manufacturers offer GNSS antennas suitable for reference/monitoring stations.

5.2.1 Redundancy

All methods of integrity monitoring rely, in one way or another, on checking the consistency of redundant information. In aviation, LAASs generally host four or five RRs in order to have redundant measurements and implement methods for fault detection. Redundancy is important to reduce integrity and continuity losses and preserve the system integrity even in case of faults that can be detected and isolated. The use of redundant RRs for the computation of differential corrections and, most important, for the integrity monitoring is a common practice used in LAAS for aviation.

5.3 Local Ground Facilities - Differential GPS

According to [3], LAAS is a local-area Differential Global Positioning System (DGPS) because it places all its reference receivers close together (within tens or hundreds of meters) and forms a single correction for each satellite. LAAS is more accurate than SBAS provided that the user has a baseline (i.e., distance between the RR and the user) lower than 100 km. One of the main operations performed by a LAAS consists in the computation of PRC and their broadcast over the geographical region covered by the LAAS broadcasting facility. Table 5.1, taken from [4], reports the error budget on the pseudorange, listing all possible error sources. The table compares the expected errors, using a stand-alone GPS receiver and Local-area Differential GPS (LADGPS) corrections. The table does not consider the error induced by the LGF on the final pseudorange estimate using DGPS.

Table 5.1 Error on the pseudorange for GPS, with and without DGPS corrections

Segment	Error source		1σ error (m)
		GPS only	DGPS
Space/control	Satellite clock	1.1	0
	P(Y)-C/A group delay	0.3	0
	Broadcast ephemeris	0.8	0.1–0.6 mm/km baseline in km
User	Ionospheric delay	7.0	0.2–4 cm/km baseline in km
	Tropospheric delay	0.2	1–4 cm/km baseline in km
	Receiver noise and resolution	0.1	0.1
	multipath	0.2	0.3
System UERE	Total	7.1	0.3 m + 1–6 cm/km baseline in km

Note that common errors at the RR and at the user receiver are mitigated and the benefit of differential corrections depends on the baselines. Errors due to local effects, such as multipath, cannot be removed. The following sections describe methods for the computation and adjustment of PRC and their broadcast.

5.3.1 Pseudorange Corrections

Each RR computes the PRC as:

$$c_{PR,n,m} = \rho_{obs,n,m} - r_{cal,n,m} - c\tau_{SV,n} \tag{5.1}$$

where we use ρ to indicate the pseudoranges, whereas the term r refers to ranges. In particular:

- $\rho_{obs,n,m}$ is the pseudorange related to the nth satellite, as observed by the mth RR. The term includes errors due to the ionosphere, troposphere, clock offset of the mth RR and noise. It is generally smoothed with carrier-phase measurements, to mitigate the effect of the thermal noise. Details on the carrier smoothing techniques can be found in [5, 6].
- $\tau_{SV,n}$ is the clock correction of the nth satellite, included in the navigation message.
- $r_{cal,n,m}$ is the geometric range respect to the nth satellite and computed by the mth RR. It is obtained through the difference of the satellites position (recovered processing ephemeris data broadcast in the navigation message) and the RRs position (known as the site is surveyed). This terms is affected by ephemeris errors and uncertainties on the surveyed position estimated during the RR installation.
- c is the speed of light.

The correction $c_{PR,n,m}$ in (5.1) is corrupted by errors affecting both the $\rho_{obs,n,m}$ and $r_{cal,n,m}$. From a pure theoretical perspective, such correction could include errors due to multipath and interfering signals that might be present at the RR antenna.

Equation (5.1) can be rewritten as:

$$c_{PR,n,m} = r_{mis,n,m} - c\tau_{RR,m} - r_{cal,n,m} - c\tau_{SV,n} \tag{5.2}$$

where we substituted the observed pseudorange (i.e.,$\rho_{obs,n,m}$) with the difference between the range (i.e., $r_{mis,n,m}$) and the error due to the RR clock offset (i.e., $\tau_{RR,m}$). Equation (5.2) can be written in a more suitable form to better appreciate the meaning of the PRC:

$$c_{PR,n,m} = \Delta r_{mis,cal}(m, n) - c\tau_{RR,m} \tag{5.3}$$

where $\Delta r_{mis,cal}(m, n)$ is the difference between the measured and the geometric range:

$$\Delta r_{mis,cal}(m, n) = r_{mis,n,m} - r_{cal,n,m} - c\tau_{SV,n} \tag{5.4}$$

If the distance between the user and the LAAS is small (i.e., short baseline), the term $c_{PR,n,m}$ contains the same iono and tropo errors that corrupt the users' measurements. Therefore, the PRC can be used to remove such errors at the users' receiver. Note that there is also another way to compute the PRCs. Ionospheric and tropospheric corrections, as well as the clock offset adjustments, recovered through the navigation massage are applied to the observed pseudoranges. This second method is equivalent to the method described above, providing that the receiver follows the same strategy, before applying the PRCs.

5.3.2 Clock Adjusted PRC

In aviation applications, the Federal Aviation Administration (FAA) specifications [7] introduce the concept of clock adjusted PRCs, which can be written as:

$$c_{PR,n,m,a} = c_{PR,n,m} - \frac{1}{N_c}\sum_{n \in S_c} c_{PR,n,m} \tag{5.5}$$

where

- the second term of (5.5) is the clock adjustment.
- S_c represents the set of satellites visible by all RRs of the LAAS and declared valid as ranging sources;
- N_c is the number of valid satellites.

The clock adjustments is a rough estimate of $\tau_{RR,m}$. The motivation for such an adjustment is clearly explained in [8]. As stated in that paper, in the computation of the user position, any bias common to all pseudorange measurements does not affect the position, but only the estimated user clock bias. Therefore, from a theoretical point of view, there is no need to remove the effects of the reference receiver clock bias from the corrections, as it is done in (5.5). However, the reference clock bias is removed as a matter of convenience. In fact, $c_{PR,n,m,a}$ can be represented with a lower number of bits, being numbers in a smaller range. This allows to conserve broadcast bandwidth and to produce continuous corrections. In turn, this yields to decrease the dynamic range of the broadcast corrections and to ensure the continuity of the corrections. Similarly, the calculated satellite clock errors are removed by the reference station to decrease the dynamic range of the corrections. The reference clock bias estimate, that is removed, does not need to be exact, since the residual reference clock bias error will be identical on all the corrections. Therefore, it will only affect the users' clock bias estimate. Two methods are proposed in [8] to remove the reference clock bias. One of these is compliant with the expression in (5.5).

Notice that in [8], authors propose to filter the second term of (5.5) prior to adding it to the corrections. Such filtered version is not present in the broadcast corrections in the FAA document [7]. The rest of the paper [8] is devoted to the description of algorithms designed to attenuate the effects of the non-common mode errors on the broadcast corrections. Notice that the removal of the RR clock bias is not present in other documents on DGPS. Kaplan [4] does not introduce any clock adjustment and introduce the concept of composite clock offset, including both RR and user clock biases.

5.3.3 Broadcast PRC

Reference [9] specifies that the broadcast PRCs shall be calculated using the equation:

$$c_{PR,n} = \frac{1}{M(n)} \sum_{m \in S_{RR,n}} c_{PR,n,m,a} \tag{5.6}$$

which is the average of the clock adjusted PRCs over $S_{RR,n}$ which is the set of RRs with valid measurements for the nth satellite. In (5.6), $M(n)$ indicates the number of RRs that are able to perform valid measurements on signal transmitted by the nth satellite (e.g., $M(5) = 3$ means that PRN 5 is received by 3 RRs).

5.3.4 PRC Standard Deviation

In order to evaluate the confidence interval limit (i.e., the PL), the user receiver needs to know the standard deviation of the error introduced by LAAS in the computation of the broadcast PRCs. This standard deviation, indicated as $\sigma_{pr,gnd}$, will contribute to the total standard deviation of the errors performed by the user at the output of the PVT subsystem. Therefore, also $\sigma_{pr,gnd}$ is broadcast by the LGF. According to [7], $\sigma_{pr,gnd}$ shall be broadcast for each ranging source, so that:

$$\sigma_{pr,gnd} = \sqrt{\sigma^2_{pr\_lgf} + \sigma^2_{spatial\_dec}} \tag{5.7}$$

where

- $\sigma^2_{spatial\_dec}$ is stored in the non-volatile memory of the LGF and takes into account the geometry of the RRs in the LAAS;
- $\sigma^2_{pr\_lgf}$ depends on $M(n)$ and on the elevation of the valid satellites used in (5.6).

Note that in many other papers, only the term $\sigma_{pr,gnd}$ is mentioned. For example in [10], authors refer to $\sigma_{pr,gnd}$ as the broadcast standard deviation of the fault-free errors in the average differential correction for the nth satellite due to such sources as RR noise and nominal multipath. According to [7], the accuracy of LGF shall be such that σ_{pr_lgf} does not overcome a given maximum value, which represents an accuracy requirement for the LGF design.

5.4 Local Ground Facilities and Integrity Monitoring

The concept of integrity introduced in Sect. 1.2 has to be extended to the operations performed by the GBAS ground stations. Moreover, a GBAS station can also implement complex algorithms of integrity monitoring to assist the GNSS equipment of the users. The following sections are devoted to the description of the method of integrity monitoring implemented in LAASs. Therefore, these methods are based on error statistical models, all derived for aeronautical applications. The adaptation of such models to other environments (e.g., road, rail) is not straightforward and deserves a careful analysis. In general, few publications (either journals, conference proceeding, and PhD thesis) can be found on integrity for non-aviation applications, and important open points seem not addressed yet. In conclusion, the methods described in this chapter have to be considered only as starting points for the design of new techniques, each one well tailored to a specific application.

5.4.1 B-Values

This section describes the method commonly used to check the integrity of a DGPS reference receiver of a LAAS.

If different receivers are installed at the reference station, ideally the PRCs from all of them should be the same for a given satellite. Therefore, the differences on PRCs across reference receivers, called *B-values*, can be conveniently used as test statistics to check the integrity of the receivers. According to [11], the B-values are the best estimate of pseudorange correction errors under the hypothesis that a given reference receiver has failed. The B-value, indicated with $B(n, m)$, is derived by subtracting two averages:

- the average PRC for the nth ranging source calculated using information from all available references,
- the average correction for the same ranging source calculated with the reference under test excluded.

From a mathematical perspective, the B-value associated with the ith receiver installed within a reference station can be written as:

$$B(n,i) = c_{PR,n} - \frac{1}{M(n)-1} \sum_{m \neq i, m \in S_n} c_{PR,n,m,a} \tag{5.8}$$

where

- $c_{PR,n}$ is the PRC defined in (5.6);
- $c_{PR,n,m,a}$ is defined as in (5.6);
- M is the number of receivers installed at the reference station;
- m is an index that goes from 1 to M;
- i indicates the receiver under test;
- n is the index for the satellites in view;
- $M(n)$ is the number of receivers able to receive the nth satellite (e.g., $M(5) = 3$ means that PRN 5 is received by 3 receivers at the reference station);
- S_n is the set of receivers with valid measurements for PRN n;
- a indicates the PRC after clock adjustment (see [7] and Sect. 5.3.2 for details).

The B-value represents the estimate of the bias in the measurement of the nth ranging source as measured by the ith reference receiver. The resulting value is compared against an integrity threshold, which is based on the continuity requirement. The ground augmentation system calculates differential corrections for an individual satellite, only if the B-values from at least two RRs are below the integrity threshold. Note that, such an error estimate and measurement exclusion are valid only if the local errors measured at the individual reference locations are independent. Correlated errors will decrease the accuracy of the estimation and can potentially degrade system integrity and continuity.

The B-values are also used to evaluate the PL in the hypothesis of a single RR fault, as described in the Sect. 5.6. For this reason, they are broadcast to the user.

5.4.2 *Analysis of Methods for Sigma-Mean Monitoring*

In the literature, methods for the monitoring of the statistical characteristics of PRC errors, namely the methods of sigma-mean monitoring, are described. Although these methods are currently proposed for augmentation systems in aviation, they might be considered in other fields of applications to further mitigate the loss of integrity. According to [11], aircrafts using LAAS corrections compute the limits on the position error in the fault-free hypothesis, assuming a zero-mean, normally distributed, fault-free error model for the broadcast PRCs. This computation is a significant part of the final PL computation, performed by the user, as described in Chap. 4. User integrity thus relies on the standard deviations of PRC errors (i.e., $\sigma_{pr,gnd}$) that are broadcast to users along with the corrections, since such standard deviations contribute to the computation of the final standard deviation used by the user in the PL computation. If the broadcast error model does not properly represent the true (unknown) error distribution, a serious threat to the aircrafts may result. Thus, special care must be taken to validate the adopted error model.

5.4.2.1 Major Sources of Statistical Uncertainty

As already mentioned in Sect. 5.3.4, the standard deviation σ_{pr_lgf} must not overcome a given maximum value, which represents an accuracy requirement for the LGF design. However, during the operative life of the facility, violations of this requirement could happen. According to [11], the main sources of statistical uncertainty, due to unexpected anomalies, are:

- site installation errors;
- non-stationary error distributions caused by multipath variation;
- receiver noise error amplifications due to any natural system failure;
- other possible malfunctions.

Notice that error sources such as ground reflected multipath and systematic reference receiver/antenna errors may be nonzero-mean Gaussian distributed. The same concept is addressed in [10], where the author states that ranging source (RS) integrity monitoring refers to the LGF monitors concerned with failures involving the satellite signal and navigation data. With respect to [10, 11] adds other error sources that potentially distort the received signal:

- Radio Frequency Interference (RFI);
- low signal level;
- code-carrier divergence;
- excessive satellite clock acceleration and ephemeris;
- others.

Thus, real-time monitoring is required to respond to failure events, where the true sigma grows to exceed the broadcast sigma or the true mean becomes substantially nonzero.

5.4.2.2 Examples of Methods for Sigma Monitoring

To validate the safety of the broadcast $\sigma_{pr,gnd}$, the LAAS must verify in real time that a Gaussian distribution with zero mean and the broadcast standard deviation overbounds the true (unknown) error distribution. Because of the direct connection between the broadcast $\sigma_{pr,gnd}$ and user integrity, real-time sigma monitoring is also necessary to detect anomalies, or failure events where, during operation, the true sigma exceeds the broadcast $\sigma_{pr,gnd}$. Several methods have been proposed to implement the sigma-mean monitoring. In [11], authors suggest to exploit the relationship between the variance of the B-values and the broadcast $\sigma_{pr,gnd}$, for the nth ranging source, which is:

$$\sigma_{B,n,i}^2 = \sigma_{B,n}^2 = \frac{\sigma_{pr,gnd}^2}{M(n) - 1} \tag{5.9}$$

Such a value can be used to normalize the B-values, so obtaining a unit-variance quantity $B_{norm}(n, i)$. Estimating the variance of $B_{norm}(n, i)$, a value of the order of one is expected. If the estimated variance deviates from one, this means that the true value of $\sigma_{pr,gnd}$ is deviating from the theoretical nominal value (broadcast to the user). Therefore, the sigma monitor system generates an alert if the estimated sigma exceeds a threshold. This method of sigma-mean monitoring belongs to the class of the Range Domain Method (RDM). Authors in [11] affirm that RDM may be conservative, since the range domain method requires a transformation from pseudorange correction errors to position error estimates. Then, an alternative technique proposed in the literature is the so-called Position Domain Method (PDM), which avoids the conservatism by performing a safety check directly in the position domain. In LAAS for aviation, PDM-based monitoring needs the deployment of a remote receiver, which emulates the aircraft conditions as much as possible. In general, this receiver represents a pseudouser with known position coordinates. The PDM collects measurements with this remote receiver and derives position solutions by applying LGF corrections to all visible satellites approved by the LGF and all possible subsets of satellites. The position solutions are then compared with the known (surveyed) location of the PDM antenna, and errors exceeding detection thresholds are alerted. According to [12], current LGFs are based on the RDM, because of the complexity of PDM monitoring systems. However, given that an enhanced LGF architecture is required to meet Category II/III requirements, PDM monitoring techniques are proposed in [12] to meet these requirements.

5.4.3 Gaussian and Non-Gaussian Error Model

In the previous sections, the PRC errors are considered continuous random variables with a zero-mean Gaussian probability density function. However, this model is not necessarily justified by experimental data. The Gaussian model is motivated by the properties of its distribution, which benefit practical implementation. In particular,

- It simplifies the communication of the error statistical characteristics, since zero-mean Gaussian distributions are described by a single parameter: the standard deviation. Therefore, the transmission of the standard deviation for each approved satellite (the so-called $\sigma_{pr,gnd}$) does not overload the bandwidth of the GBAS communication channel.

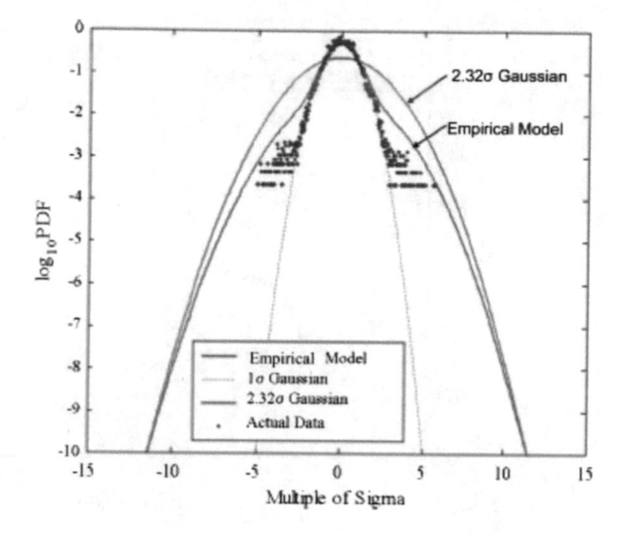

Fig. 5.2 Probability density function of normalized B-value (figure taken from [12])

- It simplifies the computation of PL, since a closed-form equation exists for this distribution, as given in (4.20).

Experiments indicate that GNSS large errors, beyond 2–3 sigma, occur with a greater than Gaussian frequency, even though nominal small and moderate errors are distributed in an essentially Gaussian manner. An interesting experiment is shown in Fig. 5.2, taken from [12], which represents the distribution of the LGF B-values, in logarithmic scale, collected, for a period of 5 h, by the Stanford Integrity Monitor Testbed (IMT), an LGF prototype designed by Stanford University. We know that the B-values are linear combinations of PRC differences, as described in the previous sections; therefore, they can be used to analyze the statistical characteristics of LGF errors, since they must exhibit a Gaussian probability density function if the LGF errors are Gaussian distributed. On the contrary, we can see in the figure that the distribution of actual data has non-Gaussian tails, while is well modeled by a Gaussian function for small and moderate data values.

Authors in [12] affirm that ground reflection multipath and systematic reference receiver/antenna errors could generate non-stationary error distributions, and that the fattened tails of the error distribution in Fig. 5.2 are due to this type of time-varying errors. This assertion can be verified with MATLAB experiments. Figure 5.3 shows the results of the experiments obtained by generating a sequence of random variables, each one composed of two elements: the first one (simulating thermal noise and diffuse multipath) is a zero-mean Gaussian random variable with $\sigma = 0.7\,\text{m}$, and the second one (simulating the effect of ground reflection multipath in the PRC error) is a sporadic element with uniform distribution in the interval $(-4.2\,\text{m}, +4.2\,\text{m})$, and randomly appearing with a probability Pr(MP)$= 10^{-3}$. The histogram of the simulated errors shows the expected heavy tails. It is clear that this is a very basic experiment, and more accurate error models have to be considered in a deeper analysis of this phenomenon.

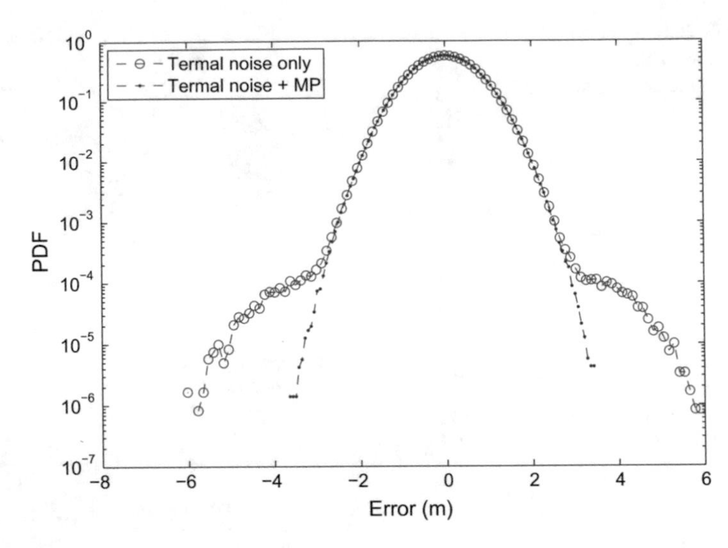

Fig. 5.3 Probability density function obtained with an experiment which emulates the generation of a random variable characterized by a distribution with heavy tails

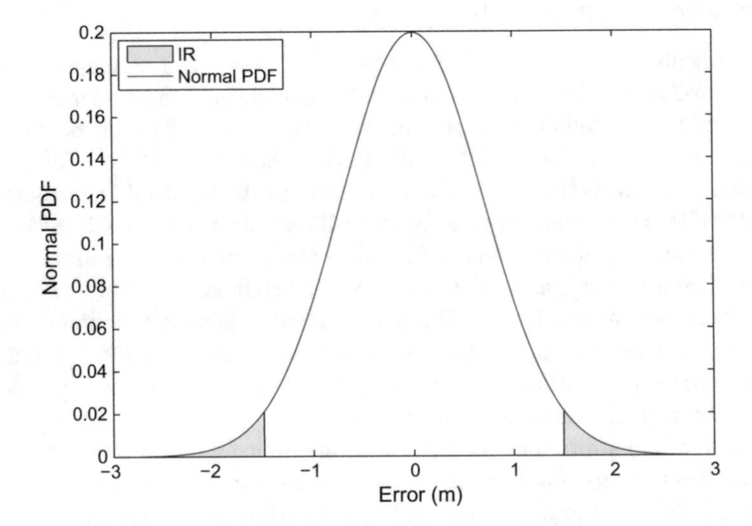

Fig. 5.4 Integrity risk definition

From these results, it is evident that the correct modeling of the distribution tails is critical, especially when integrity risk requirements are very demanding. In fact, the integrity risk depends only on the tails of the probability density function, as shown in Fig. 5.4. In these cases, countermeasures must be adopted to guarantee the correctness of the PL computation.

5.4.3.1 Gaussian Overbounding and Sigma Inflation

A method to take into account the heavy tails of the actual data distribution is to inflate the sigma, such that the broadcast distribution (characterized by $\sigma_{pr,gnd}$) overbounds all reasonable error distributions out to the probabilities assumed in the computation of the PLs. The inflation generates conservative Gaussian models, also called overbounds. This approach was proposed by Shiveley and Brass in [13], and is currently used in LAAS for integrity monitoring. The method consists in modeling the distribution of the actual data with two different functions; in particular the central part of the distribution is Gaussian modeled with a certain σ_{central}. Then an over-bound model is established which takes the form of a zero-mean Gaussian distribution with an inflated variance with respect to σ_{central}. This method has become the benchmark for the following research due to its effectiveness in practice.

A key issue of this approach is to limit the inflation factor, which could degrade the system availability if a too conservative solution is adopted. Another key problem is the capability of observing rare values during the experiments. For these reasons, a great amount of work has been done in these last ten years in this research area. An example of inflation can be seen in Fig. 5.2, taken from [12], where authors analyze the problem of sigma overbounding for LAAS and propose a method, which exploits the possibility to deploy a pseudouser station as an element of the LAAS infrastructure. Other aspects and methodologies related to this problem can be found in [14], and in the references from [31–37] reported in [15].

It is interesting to observe that this problem is becoming of interest also in other application domains. For example, in [16] authors propose overbounding techniques for road applications. They write in the abstract: "*Certain GNSS applications conceived for road users in urban scenarios must meet some particular integrity requirements to assure the system safety, reliability or credibility. For instance, GNSS-based Road User Charging is one of these applications that recently has attracted special interest. A correct design of such applications needs the knowledge of the GNSS error distribution. Furthermore, the GNSS error model should have been built with overbounding techniques. The user is a vehicle equipped with a GNSS receiver that may track different signals of various systems (GPS, Galileo, SBAS), in a single or dual-frequency configuration. The different error sources contributing to the total pseudorange error are identified, analyzed and modeled, using overbounding techniques when necessary. Finally the pseudorange measurement error model is obtained and analyzed for different receiver configurations.*"

5.4.3.2 Non-Gaussian Models

The problem of the statistical characterization of the GNSS error is still a hot topic, and also alternative models are proposed in the literature. Two examples are given hereafter.

- Rife e Pervan in [17] proposes a conservative, discrete model, called NavDEN, as a practical alternative to classical Gaussian overbounding. According to the authors, NavDEN is a particular form of a discrete error distribution, which compares favorably to Gaussian models, both in providing more margins for tail uncertainty and, at the same time, in providing generally tighter PLs when multiple distributions are convolved.
- In [18] authors propose an approach, well summarized in the abstract. They write: *Four basic error sources exist for residual pseudorange errors in a single-frequency Differential GPS system for GBAS: signal multipath, increased receiver noise (carrier-to-noise density ratios (C/N0)) due to interference, residual differential troposphere error, and the error induced by ionosphere gradients. Without restricting ourselves to classical Gaussian overbounding, we combine their probability density functions to a total pseudorange error distribution. This distribution is propagated through the GBAS Hatch filter and then mapped into the position domain using a worst case (selected by maximum VDOP) of a full 31 satellite constellation with the two most critical satellites failed observed at Braunschweig Airport, Germany. Our calculations yield a significant reduction amounting to 46% of the position domain error at the* 1.510^{-7} *integrity risk level when compared with the classical Gaussian overbounding approach.*"

5.4.3.3 Final Remarks

Although only the problem to introduce an inflated σ_{pr_gnd} has been considered in this section, it should be noted that the available methods are in principle applicable to user and residual errors as well. This is explicitly affirmed also in [14], but the corresponding analysis is not presented. This aspect should be considered and analyzed for any safety-critical applications, taking into account the environmental characteristics of the specific application.

5.4.4 GNSS Signal Quality Monitoring

Although several error sources are difficult to mitigate (e.g., thermal noise and multipath), for high-performance applications it is critical that any signal structure deviation from ideal be quantified to establish error budgets and also enable detection of minor distortions of the underlying structure from satellite-induced hardware errors. GNSS signal observation is becoming particularly important, especially within the context of integrity, which will warn the users in case the position error exceeds a predefined threshold.

Some of the earliest published results on GNSS signal quality monitoring resulted from detected anomalous behavior on satellite vehicle number (SVN) 19. Differentially corrected measurements were made, and it was found that when SVN 19 was included in the solution set, the differentially corrected vertical error could be up to

10 m [19] as opposed to the 3–8 m error when SVN 19 was not included in the solution. In order to determine the fault, signal observations were performed using a high-gain antenna. Such measurements provided a number of insights such as a distorted frequency spectrum as well as a delay in C/A code transitions on SVN 19 with respect to P(Y) transitions [20]. The anomaly became known as the *evil waveform*.

Other types of GPS evil waveforms are possible, and there is the potential for such waveforms to also occur in the signals of other GNSS systems, [21]. Over the last years, two models have been developed from the observations of the distorted signals [22]:

- the first, referred to as Evil Waveform type A (EWFA), is associated with a digital distortion, which modifies the duration of the GPS C/A code chips, as shown in Fig. 5.5. A lead/lag of the pseudorandom noise code chips is introduced. The $+1$ and 1 state durations are no longer equal, and the result is a distortion of the correlation function, inducing a bias in the pseudorange measurement equal to half the difference in the durations.
- The second model, referred to as Evil Waveform type B (EWFB), is associated with an analog distortion equivalent to a second-order filter, described by a resonance frequency and a damping factor, as depicted in Fig. 5.5. This failure results in correlation function distortions different from those induced by EWFA, but which also induces a bias in the pseudorange measurement [22].

At the time of the problem on SVN 19, it was only possible to fully observe these issues using a high-gain parabolic antenna because it guaranteed that the signal power rose above the noise floor, increasing the observability of the raw signal at the front end output. In fact, the high-gain antenna provides a positive signal-to-noise ratio (SNR), and both the individual chips and the navigation bits can be observed without performing the despreading procedure. Even today these high-gain observations are periodically performed by the US Air Force in order to ensure signal integrity via direct measurements of the signals broadcasted by GPS satellites.

Another approach proposed in the past years to investigate the GNSS signal structure is based on the analysis of the correlation function between the incoming signal and a local replica of the pseudorandom code [23]. In this case, traditional receivers have been adapted to perform specialized signal processing enabling an indirect monitoring capability in the correlation domain. Signal quality monitoring (SQM) algorithms propose a multicorrelator structure to detect possible distortions on the correlation due to irregularities on the received signal. These algorithms employ from three or more correlator pairs per channel, each slaved to the tracking pair. The measurements from each correlator output are used to form detection metrics, which are, in general, simple algebraic combinations of the measurements. Obviously, this approach provides a means to detect distortions on the transmitted signal, but it does not provide direct observation of the distortions themselves like high-gain antennas.

Figure 5.6 shows the correlation peak associated with the GPS C/A code signal tracked by a multicorrelator structure, with six correlators on the early side (i.e., 1–6), six correlators on the late side (i.e., 10–15), three correlators on the correlation peak

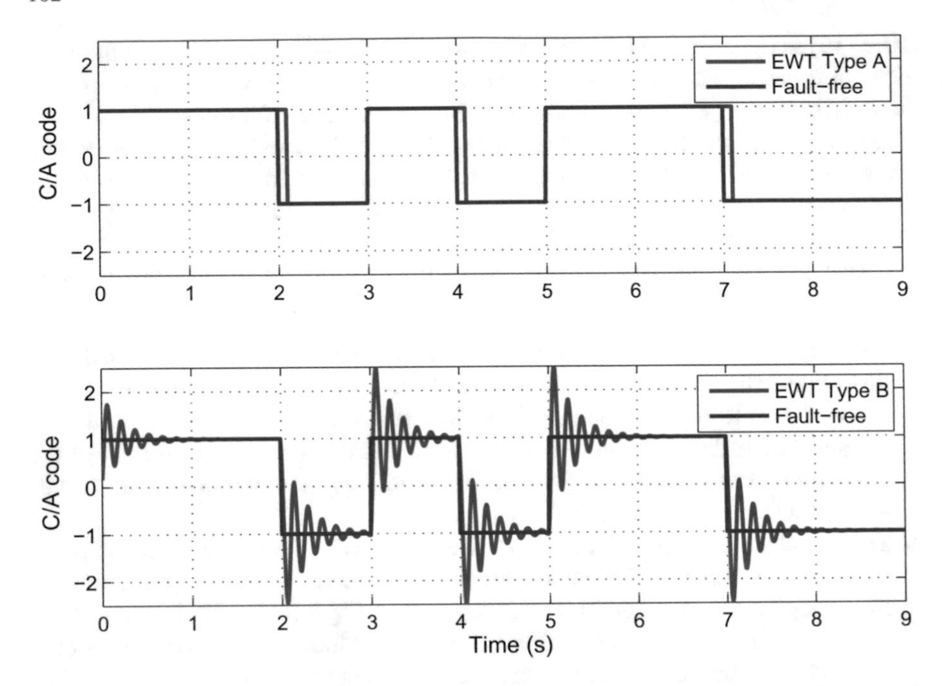

Fig. 5.5 Theoretical L1 C/A code-chip waveforms in the presence of an EWF type A (top) and EWF type B (bottom)

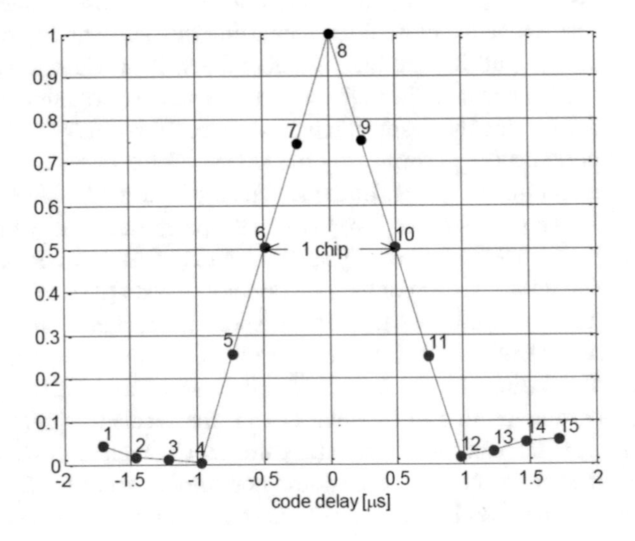

Fig. 5.6 Multicorrelator structure tracking the GPS C/A code correlation peak

(i.e., 7–9). A correlation pair, for example, correlators 6 and 10, is used for tracking purposes and control the other pairs.

Considering Fig. 5.6, at the end of each integration period (e.g., 20 ms), a set of 15 correlations can be used to form a detection metric that is normally compared against a threshold to determine whether the correlation is distorted or deviates from the nominal shape. Several metrics for SQM have been proposed in the past, some of them have been reported and compared in [24]. Two common SQM detection tests are the Delta Test and the Ratio Test. The Delta Test has been proposed to identify asymmetric correlation peaks:

$$\Delta_m = \frac{I_{E,m} - I_{L,m}}{I_{P,m}} \tag{5.10}$$

where $I_{E,m}$, $I_{L,m}$, and $I_{P,m}$ are the respective in-phase early, late, and prompt correlators, while m is an index to indicate the correlator pair in the multicorrelator structure. The division by $2I_{P,m}$ normalizes the ideal correlator peak to a maximum value of 1.

The metric associated with the Ratio Test [25, 26] is defined as:

$$\Delta_m = \frac{I_{E,m} + I_{L,m}}{2I_{P,m}} \tag{5.11}$$

The Ratio Test was originally designed to identify flat, or abnormally sharp, or elevated correlation peaks. Figure 5.7 shows the trend of the Ratio Test metric, processing a distorted GPS signals between 56,4 and 56,6 seconds (the timescale is in microseconds). The figure has been obtained in simulation, and the correlation was artificially made asymmetric for 2 s. Being the Ratio Test designed to reveal asymmetries on the correlation function, as expected, its trend became significantly irregular during the overlapping phase, when the false and the counterfeit correlation peaks collide.

It is important to note that SQM algorithms work within the receiver. In addition to the evil waveforms, they can be employed to detect signal anomalies due to the propagation environment (e.g., multipath and strong interference) or intentional spoofing attacks. The use of SQM algorithms in stand-alone receivers for robust navigation is being explored by some research groups working in GNSS, with promising results. In [27], authors used the Ratio Test for multipath detection, providing insights on the threshold settings. Finally, in [28] authors demonstrated that the Ratio Test is also a good indicator of correlation distortions due to intermediate spoofing attacks.

5.4.5 Nominal and Non-nominal Ionospheric Characteristics

GNSS signals, when traversing the ionosphere, are delayed by a quantity proportional to the Total Electron Content (TEC), which is constantly changing. Furthermore, the

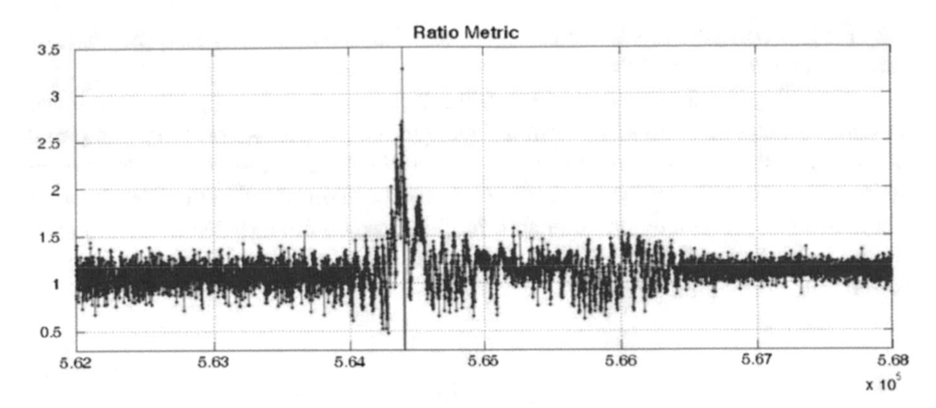

Fig. 5.7 Ratio Test metric obtained in simulation, with a distorted correlation peak from 56,4 and 56,6 seconds

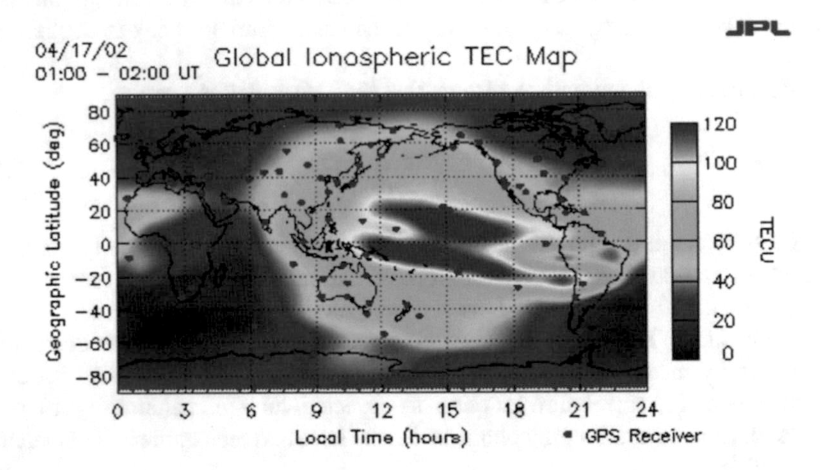

Fig. 5.8 TEC map (04/17/02)

ionospheric delay exhibits a spatial gradient, which could represent a major threat to GBAS. Nominal spatial and temporal variations in the ionospheric delays do not pose a threat in GBAS, since in this case the gradient is in the range of 2–4 mm/km (1σ), corresponding to user errors less than 10 cm. However, extremely large ionospheric spatial gradients, such as 100–500 mm/km, have been observed in the past, especially during ionosphere storm events at the time of solar maximum (i.e., in 2000–2001). Finally the gradient depends on the latitude, and low/equatorial latitude ionosphere is very volatile and characterized by intense irregularities. Figure 5.8 (taken from [29]) shows the TEC map measured on April 17, 2002, and Fig. 5.9 (taken from [30]) shows the VTEC measured on January 01, 2012. It is evident that the equatorial region is the most critical, but with time-dependent characteristics.

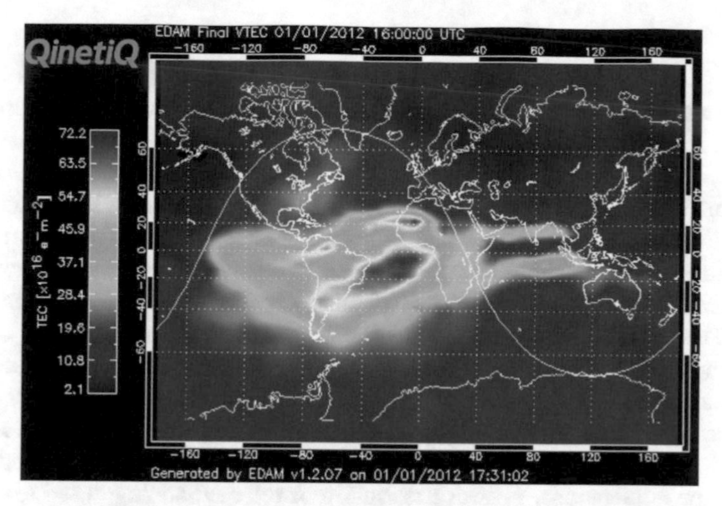

Fig. 5.9 TEC map (01/01/12)

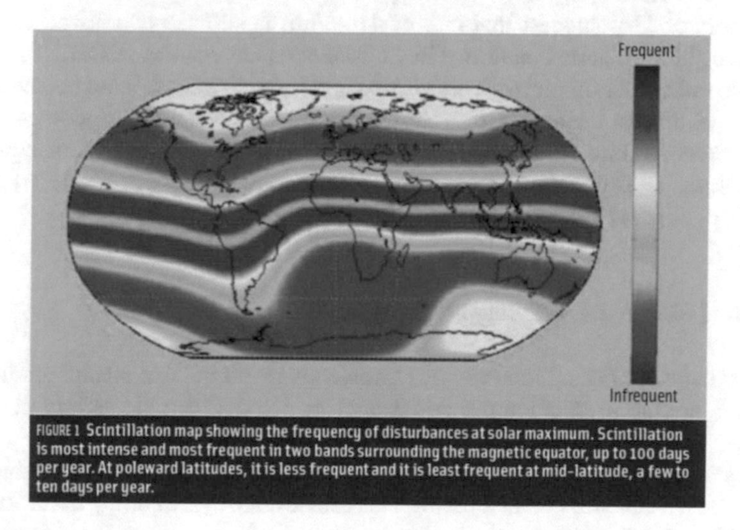

Fig. 5.10 Scintillation map

A map showing the frequency of disturbances at solar maximum is shown in Fig. 5.10, taken from [31]; again the equatorial latitude is the most critical, and recent papers [32, 33] show the increasing interest on these topics in those regions.

A LAAS must be equipped with monitors able to detect ionospheric anomalies, in particular large gradients. When using a LAAS, small residual ionospheric errors are included in the nominal conditions and are modeled in terms of standard deviation σ_{iono} in the PL computation, so modifying the variance given in (4.33). Section 5.4.5.1 gives some details on how to compute σ_{iono}. Large gradients have to be detected by GBAS, as described in Sect. 5.4.5.2.

5.4.5.1 Ionosphere Spatial Decorrelation in Nominal Conditions

It is known that PRCs' broadcast by LGF allows the removal of almost all user ionospheric error. However, residual correction errors remain due to ionosphere spatial and temporal decorrelation between RRs and user. The temporal variation is negligible during transmission time, while the spatial decorrelation is the larger of the two, and has to be taken into account when the user computes PL. This is obtained by using the broadcast value of the *vertical ionosphere gradient* (called σ_{vig}), that expresses the typical standard deviation of the ionospheric delay per user-to-reference separation. The value of σ_{iono} to be used by the user in the PL computation is obtained by a complex formula, which includes σ_{vig}, and many other coefficients, such as the horizontal speed of the aircraft, the satellite elevation, the distance between LGF and aircraft, the time constant of the smoothing filter, and others. This topic is quite complex and cannot be summarized in few lines. More details can be found in [32] for aviation applications. A clear description of the methodology used to estimate σ_{vig} can be found in [34].

Authors of [34] suggest that σ_{vig} of 4 mm/km is sufficient to cover almost all non-stormy ionospheric conditions in CONtiguous US regions (CONUS), with an adequate safety margin for more active days and for non-Gaussian tail behavior. This implies that this parameter has to be chosen, taking into account the region of interest. In fact, CONUS Iono Threat Model may be not applicable to low latitude/equatorial region. Some considerations for Australia can be found in the document in [35]. An analysis performed in Germany can be found in [36].

5.4.5.2 Monitoring of Ionospheric Anomalies

Unusual behavior during ionospheric storms may result in large spatial gradients of up to 400mm/km in slant ionospheric delay (see Fig. 5.11 for the definition of slant delay).

These anomalies have to be detected by GBAS as quickly as possible, subject to a required low probability of false alarm. This can be done by exploiting the fact that the ionosphere affects GPS signal propagation by delaying code-phase measurements, while advancing carrier-phase measurements.

This phenomenon, known as code-carrier divergence (CCD), can be described by modeling the pseudorange and carrier-phase measurements as

$$\rho_i = r_i + I_i + M_i + N_i \tag{5.12}$$

and

$$\varphi_i = r_i - I_i + b \tag{5.13}$$

where r_i is the range between the ith satellite and the aircraft, including the common mode errors such as the satellite and receiver clock offsets and tropospheric delay, I_i is the ionospheric delay, M_i is the code multipath error, N_i is the receiver noise,

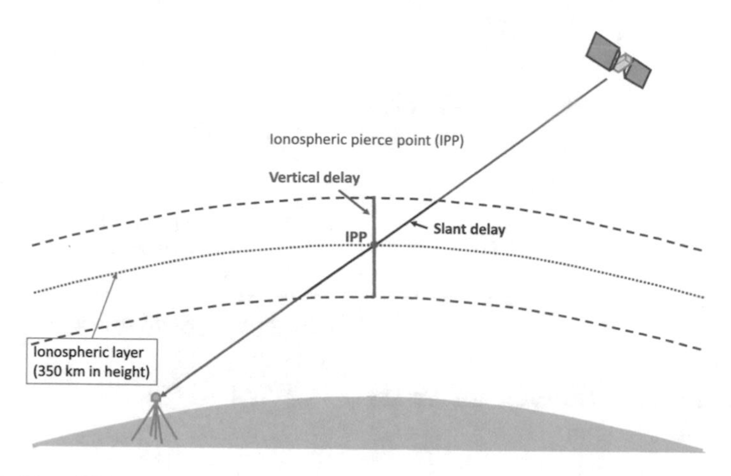

Fig. 5.11 Slant delay

and b is the phase ambiguity. The multipath and receiver noise of carrier-phase measurements can be omitted, since it is negligible with respect to those of the code measurements. A measure of CCD can be introduced, related to the difference

$$d_i = (\rho_i - \varphi_i) - (\rho_{i-1} - \varphi_{i-1}) = 2(I_i - I_{i-1}) + M_i - M_{i-1} + N_i - N_{i-1} \tag{5.14}$$

which depends on $2(I_i - I_{i-1})$ and on noise and multipath terms, which can be mitigated by filtering. Therefore, a test based on CCD can be used to detect ionospheric storms and ensure that the divergence of code and carrier for any given satellite is sufficiently small. A CCD monitor generally consists of two components: a divergence rate estimator and a detection test. The divergence rate estimator evaluates d_i (or a parameter proportional to d_i) and filters it to reduce the code noise. The filtered quantity is used as a test statistic to be compared against a threshold:

$$T_{\text{CCD}} = K_{\text{ffd,mon}}\sigma_d \tag{5.15}$$

where σ_d is the fault-free standard deviation of the test statistic and $K_{\text{ffd,mon}}$ is a constant chosen to ensure that the probability of fault-free alarm meets an allocated continuity requirement for the monitor. Other variants of this monitoring method exist, [37], and also other methods [38, 39]. Notice that in the literature, airborne CCD monitors are also proposed (see, e.g., [39]).

Finally, we report a model, introduced in 2005, which is widely used to study the impact of a large gradient on GBAS user. With this model, the ionospheric anomalies can be represented as sharp wave fronts in which a linear change in vertical ionospheric delay occurs over a short horizontal distance [40]. This linear model, shown in Fig. 5.12, is characterized by the parameters:

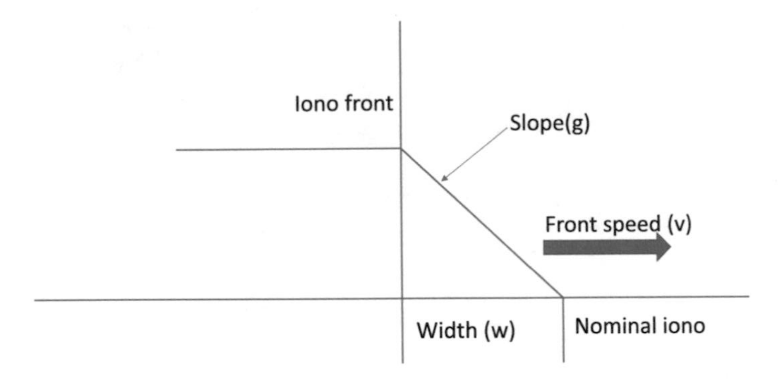

Fig. 5.12 Linear model for ionosphere anomalies

- the spatial gradient (g) in slant ionosphere delay between maximum and minimum delays;
- the width (w) of the linear change in delay;
- the forward propagation speed (v) of the wave front relative to the ground.

The upper and lower bounds on the model parameters were determined by analyzing a large set of GPS data collected from the mid-latitude sites in 2000–2004.

5.4.6 Nominal and Non-nominal Ephemeris Error

Differential GNSS removes the common mode pseudorange errors affecting both reference and user receivers. Ephemeris errors, i.e., the errors in describing the true orbit of a GPS satellite, are not perfectly common mode because the satellite orbit error components onto the line of sight from any receiver varies with receiver location. This non-common mode effect is an example of spatial decorrelation also relevant to other pseudorange errors (e.g., ionospheric and tropospheric). A bounding (worst case) residual Differential GNSS (DGNSS) ephemeris pseudorange error ε_{eph} is, [15]:

$$\varepsilon_{eph} = \frac{\|b\| \delta R}{R} \tag{5.16}$$

where

- R is the distance between the reference receiver and the given satellite (the distance from user receiver to the given satellite is essentially the same);
- b is the baseline vector (i.e., the vector between reference and user receivers);
- δR is the 3D orbit error of the given satellite.

Under nominal conditions, the GNSS satellite ephemeris errors are typically 10 m or less in three dimensions (3D). Assuming $\|b\| = 10\,\text{km}$, and a value of $R \cong 20,000\,\text{km}$, a 3D nominal satellite orbit error of 10 m results into a worst-case DGNSS ephemeris pseudorange error of only 0.005 m. This error does not pose any threats to LAAS. However, the nominal conditions do not preclude the possibility that a failure will cause satellite ephemeris errors large enough to threaten LAAS. Referring to GPS experience, no significant ephemeris failure occurred until the serious failure of April 10, 2007, which is illustrated in [15]. In this event, the GPS operational control segment deliberately maneuvered satellite SV-54/PRN-18 from one orbit to another while the SV was still flagged as "healthy". The figures in [15] show that the range error during the unflagged maneuver grew to about 350 m. Ephemeris monitoring is needed to detect and exclude the affected SV, and a relevant protection level shall be calculated to cover the remaining undetected ephemeris failures to check that integrity is met in both nominal and faulted conditions. Other details on how to take into account the ephemeris errors in th PL computation can be found in Sect. 5.6.

5.4.6.1 Ephemeris Failure Models

GNSS ephemeris failures are categorized in two types, A and B, based upon whether or not the failure is associated with a satellite maneuver (i.e., orbit change).

Type A failures (associated with satellite maneuver). Type A failures are further subdivided into two separate classes: A1 and A2.

- A1 scenario (failure after the maneuver) In the A1 scenario, the satellite maneuver is scheduled and intentional, and the SV flag is set to unhealthy by the orbit control subsystem (OCS). After the maneuver is completed, the flag is reset to healthy followed by the broadcast of erroneous ephemeris data.
- A2 scenario (failure during the maneuver) The type A2 scenario can be defined as a failure creating a hazard during the maneuver period. This is further subdivided into Type A2a and Type A2b. The failure Type A2a is a planned maneuver during which the SV health flag remains set to healthy (e.g., the incident reported in [15]). The failure Type A2b is an unintentional maneuver caused by uncommanded firing of large SV thrusters. This event is extremely improbable because it cannot be triggered automatically and because multiple failures would have to occur on the satellite.
  Type A2b is a very small subset of all SV maneuvers and are considered not credible for CAT I LAAS. The Type A2b is being addressed for CAT II/III LAAS.

Type B failures (no satellite maneuvers or orbit change). Type B failures, considered to be more credible than Type A but still very rare, include all cases where no satellite maneuvers are involved. The SV broadcasts anomalous ephemeris that produces large errors in satellite positions. This event would most likely be caused either by an error in computing the broadcast ephemeris parameters or by corruption of the correct parameters somewhere along the line from OCS creation to OCS uplink to satellite broadcast. The probability of type B failures is relatively higher than type A because orbit ephemeris uploads and broadcast ephemeris changeovers are frequent (once every two hours), whereas SV maneuvers are rare (no more than once or twice intentional maneuvers per year). In addition, the unintentional maneuvers (Type A2b) are a very small subset of all SV maneuvers and are considered not credible for CAT I LAAS. The Type A2b is being addressed for CAT II/III LAAS.

5.4.6.2 Ephemeris Failure Mitigation

Should ephemeris failures occur, the responsibility for detecting and excluding these failures would lie with the LAAS ground facility rather than with users. This is achieved by ephemeris monitoring. The probability of missed detection is covered by LAAS notifying the user with the parameters relevant to the possible (undetectable) hazard so that the ephemeris protection level can be calculated by the user to check that the integrity risk is at an acceptable level. The ephemeris failure Missed Detection is not a problem for wide-area DGNSS because the regional spread of reference stations makes it possible to directly solve for satellite positions in space independently of the broadcast ephemeris messages [15]. However, for LAAS, with reference receivers in only one location, satellite ephemeris monitoring is more difficult.

Type A Failure mitigation. Mitigation may be achieved by proposed monitors that difference the computed and measured ranges and range rates and compare the resulting test statistics to predefined thresholds, which are based on the fault-free behavior of the test statistics. The effectiveness of these monitors are analyzed and verified in [41]. The proposed monitors meet the CAT I precision approach integrity risk allocations to these threats [41].
For example, in LAAS, the maximum pseudorange correction that can be broadcast is 327.67 m [15], but a much smaller threshold of ± 125 m can be applied to this correction without sacrificing continuity or availability. A threshold of ± 3.4 m/s can be applied to the range-rate correction, and a threshold of $0.019\,\text{m/s}^2$ can be applied to the estimated range acceleration. These thresholds will detect almost all satellite maneuvers, whether commanded or uncommanded by OCS, but there is a small region of worst-case maneuver geometry that could go undetected. Thus, this approach does not cover all possible threatening geometries. Its intent is to reduce the space of possible hazards so that the overall integrity requirement is met [15].
An alternative differential-carrier-phase-based algorithm proposed for detecting "Type A1" failures is described in [15]. This method makes the SV maneuvers observable by utilizing the same ephemeris error decorrelation effect that creates a potential threat to users. Carrier-phase differences are used to make these effects visible over reference receiver antenna baselines of 100 m to 1 km [15].
Type B Failures mitigation. To mitigate Type B threats, the LGF stores validated ephemeris from previous days and uses them to project forward an independent predictive estimate of the current estimate for comparison. The ephemeris Type B monitor is described in detail in [42]. The Type B monitor is not effective against the Type A failure because the predictive capability of the monitor is compromised

by the intervening maneuver. Several monitors that are useful in detecting type B ephemeris failures within local-area DGNSS are described in [42]. These methods are based on the "YE-TE" concept, which compares the "yesterdays ephemeris" (stored from the most recent pass of a given satellite) to the "todays ephemeris", meaning the ephemeris that is broadcast by a newly risen satellite at the present time. From [42], at any epoch k the difference between r_k, the satellite computed range from current ephemeris (TE), and $\hat{r}_k$, satellite computed range from prior estimated ephemeris (YE), gives the position deviation vector

$$\delta r_k = r_k - \hat{r}_k \tag{5.17}$$

From δr_k it is possible to build a scalar test statistic s_k, with a chi-square statistical distribution, that will be compared to a threshold T to detect an anomalous ephemeris. The threshold T is defined to ensure a fault-free alarm probability P_{FFA} consistent with system availability and continuity requirements. Given an ephemeris anomaly, the test statistic s_k will have a Non-central Chi square distribution around a non-centrality parameter (bias λ), which is a function of the actual ephemeris error. Using an approach similar to the one used in Sect. 4.1.2.2 to introduce the concept of slope, it is possible to define a minimum detectable ephemeris error (MDE). This quantity determines the non-centrality parameter λ that, taking into account the threshold T, results in the required probability of Missed Detection P_{MD}. More details can be found in [43].

Referring to Sect. 5.6, the parameter $S_{n,\text{vert}}$ in (5.21) is a function of MDE. The LAAS broadcasts to the user the so-called decorrelation parameter P-value for each satellite, defined as:

$$P_j = \frac{\text{MDE}_j}{\rho_j} \tag{5.18}$$

allowing the user to calculate the vertical ephemeris protection level.

5.4.7 *Other Integrity Monitoring Subsystems*

The integrity monitoring systems examined in the previous sections do not complete the list of all possible blocks included in the LAAS architecture. For example in IMT, i.e., the LAAS prototype in Stanford, the following additional monitors are present:

- Data Quality Monitoring (DQM): detects anomalies in satellite navigation data (ephemeris and clock data) to verify that satellite navigation data is of sufficient fidelity, for each satellite that rises into view of the LGF.
- Measurement Quality Monitoring (MQM): confirms the consistency of the pseudorange and carrier-phase measurements over the last few epochs to detect sudden step and any other impulsive errors due to GPS SIS clock anomalies or LGF receiver failures.

5.5 User Receivers

GNSS receivers for aviation are able to provide accurate measurements and are generally enabled to receive signals from wide-area augmentation system (WAAS) and GBAS. In this last case, they receive PRC and integrity data over a VHF channel. DGPS is necessary to provide the performance required for vertically guided approaches. In aviation, GNSS receivers also incorporate data links to enable the transmission of the aircraft location to air traffic control.

5.6 Position Confidence Estimate at Receiver Level

In this section, we describe how the PL is evaluated in aviation applications, taking into account the availability of GBAS data. It is evident that this methodology can be adopted also in other application scenarios, but the single equations have to be likely reformulated.

The fault tree for CAT I LAAS can be found in [43], where the sub-allocation of the probability of LOI is indicated, together with the intermediate events considered for the computation of the partial PLs, that is the multiple hypotheses introduced in Sect. 4.2.3 to compute the final PL. In this fault, tailored to the vertical error, only the following intermediate events are considered:

- Nominal condition (hypothesis $\mathscr{H}_0$);
- Single-reference-receiver fault (hypothesis $\mathscr{H}_{RR}$);
- Single-satellite ephemeris fault (hypothesis $\mathscr{H}_e$).

Only the vertical error is considered, since the satellite constellation geometries result in tighter error bounds in the horizontal direction than the vertical direction. Therefore, it is sufficient to bound only the projection of errors in the vertical using a VPL, which is compared to a corresponding Vertical Alert Limit (VAL).

As written in [44], conventional GBAS avionics evaluate a fault-free bound (VPL($\mathscr{H}_0$)) and two types of fault-scenario bounds covering the case of a single ground-system reference receiver fault (VPL($\mathscr{H}_{RR,m}$)) and the case of a satellite ephemeris fault (VPL($\mathscr{H}_{e,n}$)). In conventional GBAS, users do not evaluate bounds for other types of system hazards including failures of integrity monitors to detect signal deformation, satellite clock acceleration, satellite code-carrier divergence, or anomalous ionosphere gradients. Users must rely on the LGF to bound these scenarios. LAAS users compute in real time:

- VPL($\mathscr{H}_0$) (one equation);
- VPL($\mathscr{H}_{RR,m}$) (one equation per each RR);
- VPL($\mathscr{H}_{e,n}$) (one equation per each SV).

and warning is issued (and operation may be aborted) if maximum VPL over all equations exceeds VAL. Absent an actual anomaly, VPL($\mathscr{H}_0$) is usually the largest.

Fault modes that do not have VPLs must be detected and excluded such that VPL($\mathscr{H}_0$) bounds. Residual probability that VPL($\mathscr{H}_0$) does not bound must fall within the VPL($\mathscr{H}_{nc}$) ("not covered") LAAS integrity sub-allocation. In [43] the partial VPLs are evaluated by using an equation of the type of (4.30), where $|\mu_{m,j}| = 0$ in the case of hypothesis $\mathscr{H}_0$, and $|\mu_{m,j}| \neq 0$ otherwise. In particular:

$$\text{VPL}(\mathscr{H}_0) = k_{ffmd}\sigma_{\text{vert},\mathscr{H}_0} \tag{5.19}$$

$$\text{VPL}(\mathscr{H}_{RR,m}) = |B_{m,\text{vert}}| + k_{md}\sigma_{\text{vert},\mathscr{H}_{RR,m}} \tag{5.20}$$

$$\text{VPL}(\mathscr{H}_{e,n}) = S_{n,\text{vert}} + k_{md,e}\sigma_{\text{vert},\mathscr{H}_{e,n}} \tag{5.21}$$

where

- The terms of the type σ_j are the standard deviations along the vertical axis;
- $B_{m,\text{vert}}$ is a bias due to a single RR fault (which can be obtained from the B-values, broadcast by LAAS to the user);
- $S_{n,\text{vert}}$ is a bias due to ephemeris faults;
- The k-values can be obtained by the inverse erfc function (e.g., by using an equation of the type of (4.30))

Obviously, at this point a detailed discussion on how to obtain the above described terms would be necessary, but it would require a very wide and detailed discussion, and anyhow limited to the example of the CAT I LAAS fault tree.

In brief, what is important to recall is the methodology that can be summarized as follows:

- The IR value (or LOI probability) is specified for each application;
- The fault tree is built identifying all the potentially dangerous faults;
- The LOI probability is sub-allocated;
- The partial PLs are evaluated;
- The maximum PL is selected.

The obtained fault tree is oriented to the PL evaluation and concerns the missed detected fault. In a complete GBAS system, it is required to have blocks for the exclusion of the faults not included in the PL-oriented fault tree.

5.6.1 Prior Probabilities

Prior probabilities of potentially threatening failures and anomalies are needed to complete fault tree allocation and verification. The k-values in the fault-mode PL equations are derived based on estimated prior probabilities (for satellites) or required prior probabilities (for ground equipment). Notice that the prior probabilities for satellites failure and atmospheric anomaly are beyond designers control and must be estimated, while prior probabilities for ground equipment must meet given requirements. In [43] SV prior failure probability for LAAS integrity analyses is conservatively set to 10^{-4} per SV per hour (or 4.2 10^{-6} per SV per approach).

Faults and anomalies are rare events that are often difficult to characterize by theory or data. For example, anomalous signal deformation has only been observed once, on GPS SVN 19 in 1993. Imperfect knowledge of rare events requires that (conservative) assumptions be made to make modeling and mitigation practical. Assumptions like these are often called assertions. The degree of justification for a given assertion varies with its reasonableness and its criticality. Anyway the use of assertions must not to be abused. For applications different from aviation, the prior probabilities of the basic events will have to be carefully determined.

5.7 Considerations Related to the Railway Applications

Railways have already introduced satellite-based localization systems for non-safety related applications, and, only recently, the first freight SIL 4 (i.e., safe) signaling system based on ERTMS and the satellite localization has been delivered [45, 46]. Driven by economic reasons, the use of these systems for new services and, in particular, their introduction in signaling system is seriously investigated today and tested all around the world, [47–50]. However, the introduction of these techniques in the railway sector as a standard and interoperable solution, and, in particular, in the signaling applications, is relatively slow, due to the need of getting the consensus of all the railway stakeholders such as ERA, European Economic Interest Grouping (EEIG), UNIFE [51], Union Industry of Signaling (UNISIG), and Community of European Railway and Infrastructure Companies (CER).

We have seen that today four criteria are used to assess the performance of a GNSS-based navigation system for SoL applications: availability, accuracy, continuity, and integrity. These performance indicators, mainly driven by aviation and specified by International Civil Aviation Organization (ICAO), do not have direct correspondence to railway requirement criteria, even if similar criteria have to be eventually defined and accepted by the railway community. However, some preliminary and significant works on this specific topic are already present in the literature. For example in 2003, integrity equations for safe train positioning have been developed and proposed by [52]. A way to use the concept of integrity and accuracy, and the mechanism of FDE used in RAIM is considered in [53] for railway applications. The problem related to the local integrity threats, typical of land transportation, such as multipath and masking phenomena, is addressed in [54–57].

Several European projects funded researches to explore and promote the use of GNSS in railways signaling, since the beginning of 2000s. For example, the Railway User Navigation Equipment (RUNE) project, [58], investigated the use of GNSS Integrity and Safety of Life service characteristics for defining a satellite-based system to perform train location for safe railway applications. In particular a technical solution based on GNSS receivers, the use of European Geostationary Navigation Overlay System (EGNOS), and the integration with inertial sensors and on-board odometers has been analyzed. The ESA 3InSat project [48] developed and verified a satellite-based platform to be integrated into an ERTMS system to provide the

SIL 4 (i.e., safe) train position function. This satellite-supported solution is based on the concept of Virtual Balise and on the use of a Track Local-Area Augmentation Network suitable for the railway environment. An intensive validation campaign was carried on a Railway Test Site implemented in Sardinia on a 50-km line. The GSA ERSAT-EAV project [59] aimed to verify the suitability of European Global Navigation Satellite System (EGNSS) as the enabler of cost-efficient and economically sustainable ERTMS signaling solution for safety railway applications. This objective was reached by performing measurements under real operating conditions with the support of simulation tools and models, and by defining and developing a system solution verified in the laboratory and on the field. On the other end, the main objective of the ESA SBS project [60] was the feasibility study to determine the technical feasibility and economic viability of a space-aided ERTMS railway signaling system including positioning solutions using GNSS (in combination with other onboard sensors) and wireless communications (satellite in combination with terrestrial communications).

Solutions based on augmentation systems have been also investigated, such as in [61–64], and many other projects worked in similar areas, as mentioned in [47]. However, several challenges have to be still faced, as it will be clear after reading Chaps. 6 and 7, devoted to the problems related to train positioning and navigation.

5.8 Conclusions

In this chapter, methods of integrity monitoring and PL computation tailored to aviation scenarios have been described. In non-aviation transportation applications, the concept of GBAS-augmented positioning can be also applied, but the whole system has to be adapted to the different environment. In this case, also the threats close to the user, and undetected by the GBAS ground stations, could contribute to the loss of integrity. Therefore, a deep analysis should be performed to analyze the local errors, and the possibility to design user-borne integrity monitoring blocks has to be investigated.

References

1. European Aviation Safety Agency (2013) Air operations commercial air transport. http://www.eraa.org/system/files/Air%20OPS%20CAT%20Hard%20and%20Soft.pdf
2. http://www.novatel.com/assets/Documents/Papers/cma-4048.pdf
3. Enge P (1999) Local area augmentation of GPS for the precision approach of aircraft. Proc IEEE 87(1):111–132
4. Kaplan E, Hegarty C (2006) Undestanding GPS: principles and applications, 2nd edn. Artech House
5. Misra P, Enge P (2006) Global positioning system. Signal measurements and performance. Ganga-Jamuna Press, Lincoln

6. Petovello M, Presti LL, Visintin M (2016) Can you list all the properties of the carrier-smoothing filter? INSIDE GNSS 10(4)
7. FAA Faa-e-2937, 21 Sept 1999
8. Farrell J, Givargis T (2000) Differential GPS reference station algorithm-design and analysis. IEEE Trans Control Syst Technol 8(3):519–531 (2000)
9. Perepetchai V (2000) Global positioning system receiver autonomous integrity monitoring. Master's thesis, School of Computer Science, McGill University, Montreal
10. Braff R (2001) LAAS performance for terminal area navigation. Work tech papers, Mitre
11. Lee J, Pullen S, Enge P (2006) Sigma-mean monitoring for the local area augmentation of GPS. IEEE Trans Aerosp Electron Syst 42(2):625–635
12. Lee J, Pullen S, Enge P (2009) Sigma overbounding using a position domain method for the local area augmentaion of GPS. IEEE Trans Aerosp Electron Syst 45(4):1262–1274
13. Shively C, Braff R (2000) An overbound concept for pseudorange error from the LAAS ground facility. In: Proceedings of IAIN world congress/ION 56th annual meeting, San Diego, CA, pp. 661–671, 26–28 June 2000
14. Pervan B, Sayim I (2001) Sigma inflation for the local area augmentation of GPS. IEEE Trans Aerosp Electron Syst 37(4):13011311
15. Gleason S (2009) GNSS applications and methods - GNSS technology and applications. Artech House
16. Salos D, Macabiau C, Martineau A, Bonhoure B, Kubrak D (2010) Nominal GNSS pseudorange measurement model for vehicular urban applications. In: Position location and navigation symposium (PLANS), pp 806 – 815
17. Rife J, Pervan B (2012) Overbounding revisited: discrete error-distribution modeling for safety-critical GPS navigation. IEEE Trans Aerosp Electron Syst 48(2)
18. Dautermann T, Mayer C, Antreich F, Konovaltsev A, Belabbas B, Kalberer U (2012) Non-gaussian error modeling for GBAS integrity assessment. IEEE Trans Aerosp Electron Syst 48(1):693–706
19. Kalafus R (1993) A new error source in differential GNSS operations. Technical report, Trimble Navigation
20. Edgar C et al (1999) A co-operative anomaly-resolution on PRN-19. In: Proceedings of 12th international technical meeting of the satellite division of the institute of navigation, Nashville
21. Pagot J-B (2016) Modeling and monitoring of new GNSS Signal distortions in the context of civil aviation. Ph.d. dissertation, Institut Nationale Polytechnique de Toulouse (INP Toulouse)
22. http://gpsworld.com/innovation-evil-waveforms-generating-distorted-gnss-signals-using-a-signalsimulator/
23. Phelts RE, Akos DM, Enge P (2000) Robust signal quality monitoring and detection of evil waveform. In: Proceedings of ION GPS 2000, 13th international technical meeting of the satellite division of the institute of navigation, Salt Lake City
24. Phelts RE, Walter T, Enge P (2009) Toward real-time SQM fro WAAS: improved detection techniques. In: Proceedings of ION GNSS conference in Portland
25. Phelts RE, Akos DM, Enge P (2000) Robust signal quality monitoring and detection of evil waveforms. In: Proceedings of ION GPS 2000, Salt Lake City, Utah
26. Phelts RE (2001) Multicorrelator techniques for robust mitigation of threats to GPS signal quality. Ph.D. dissertation, Standford University, Palo Alto, California
27. Fantino M et al (2009) Signal quality monitoring: Correlation mask based on ratio test metrics for multipath detection. In: Proceedings of international global navigation satellite systems society, IGNSS symposium 2009 surfers. Paradise, Australia, 1–3 December 2009
28. Ledvina B, Bencze W, Galusha B, Miller I (2010) An in-line anti-spoofing device for legacy civlil GPS receivers. In: Institute of Navigation ITM Conference, San Diego, CA, 26 Jan 2010
29. JPL. https://iono.jpl.nasa.gov/gim.html
30. European Space Weather Portal. http://spaceweather.eu/swwt/ionosphere
31. Kintner P, Humphreys T, Hinks J (2009) GNSS and ionospheric scintillation. Inside GNSS
32. Satya Srinivas V, Sarma AD, Supraja Reddy A, Krishna Reddy D (2014) Investigation of the effect of ionospheric gradients on GPS signals in the context of LAAS. Prog Electromag Res B 57:191–205

33. Miguel Juan J, Sanz J, Prieto R, Schlueter S (2013) Ionospheric activity in the South East Asian region. In: ICSANE 2013 international conference on space, Aeronautical and navigational electronics 2013, Hanoi, Vietnam. 2 Dec 2013
34. Lee J, Pullen S, Datta-Barua S, Enge P (2007) Assessment of ionosphere spatial decorrelation for global positioning system-based aircraft landing systems. J Aircr 44(5)
35. Indonesia (2010) Ionosphere characterization in asutralia to support GBAS implementation. Fourteen meeting of the communication/navigation/surveillance and meterorology sub-group of Apanpirg (CNS/MET SG/14), Jakarta, 19 July 22
36. Mayer C, Belabbas B, Jakowski N, Meurer M, Dunkel W (2009) Ionosphere threat space model assessment for GBAS. In: ION GNSS 2009, Savannah, GA, USA, 22–25 Sept 2009
37. Simili DV, Pervan B (2006) Code-carrier divergence monitoring for the GPS local area augmentation system. In: IEEE/ION position, location, and navigation symposium
38. Xie G, Pullen S, Luo M, Enge P (2009) Detecting ionospheric gradients with the cumulative sum (CUSUM) method. In: 21st international communications satellite systems conference and exhibit, Yokohama, Japan
39. Ming L, Sam P, Jed D, Hiroyuki K, Gang X, Todd W, Enge P, DattaBarua S, Dehel T (2003) LAAS ionosphere spatial gradient threat model and impact of LGF and airborne monitoring. ION GPS/GNSS, Portland, OR, 9–12 Sept 2003
40. Pullen S, Enge P (2007) An overview of GBAS integrity monitoring with a focus on ionospheric spatial anomalies. Indian J Radio Space Phy 36:249–260
41. Tang H, Pullen S, Enge P, Gratton L, Pervan B, Brenner M, Scheitlin J, Kline P (2010) Ephemeris type a fault analysis and mitigation for LAAS. In: IEEE/ION position location and navigation symposium (PLANS), Indian Wells, CA, USA, pp 654–666, 4–6 May 2010
42. Pervan B, Gratton L (2005) Orbit ephemeris monitors for local area differential GPS. IEEE Trans Aerosp Electron Syst 41(2): 449–460
43. Pullen S (2001) Tutorial presentation: augmented GNSS: fundamentals and keys to integrity and continuity. http://www-leland.stanford.edu
44. Rife J, Pullen S, Enge P (2009) Evaluating fault-mode protection levels at the aircraft in category III LAAS. In: Proceedings of the 63rd annual meeting of the institute of navigation, Cambridge, MA, pp 356–371
45. Ansaldo STS (2017) Roy hill signalling & communications - system generic application safety case verification report, Doc. Number 000091.R11.EN Rev. 06.00, 27 July 2017
46. Ansaldo STS (2017) Satellite assisted railway application - LDS generic product and application safety case, Doc. Number P60A.0100001.A01.07EN Rev. 05.00, 01 June 2017
47. Marais J, Beugin J, Berbineau M (2017) A survey of GNSS-based research and developments for the european railway signalinog. IEEE Trans Intell Transp Syst PP(99):1–17
48. ESA. https://business.esa.int/projects/3insat
49. Ansaldo STS. https://medsalt.files.wordpress.com/2016/09/10-ricercainnov-ita-mar-2017.pdf
50. FS news. http://www.fsnews.it/fsn/Sala-stampa/Cartelle-stampa/ERSAT-EAV-conclusi-test-tecnologie-satellitari-traffico-ferroviario-regionale
51. UNIFE. http://www.unife.org/
52. Nikiforov Igor V, Choquette Franois (2003) Integrity equations for safe train positioning using GNSS. Atti dell'Istit Ital di Navig 171:52–77
53. Zhu N, Marais J, Btaille D, Berbineau M (2017) Evaluation and comparison of gnss navigation algorithms including fde for urban transport applications. In: ION international technical meeting, Monterey, United States
54. Legrand C, Beugin J, Conrard B, El-Miloudi E-K (2015) Approach for evaluating the safety of a satellite-based train localisation system through the extended integrity concept. In: European safety and reliability conference, Zürich, Switzerland
55. Wendel J, Schubert F, Floch J-J, Ioannides R, Wullems C (2016) GNSS-based integrity for railway users using map-aided solution separation. NAVITEC ESA/ESTEC, The Netherlands, 14–16 Dec 2016

56. Schubert F, Gulie I, Wendel J, Wullems C, Ioannides R, Kohl R (2016) A geometrical-statistical multipath propagation model for railway navigation applications. NAVITEC ESA/ESTEC, The Netherlands, 14–16 Dec 2016
57. Grosch A, Crespillo OG, Martini I, Gnther C (2017) Snapshot residual and Kalman filter based fault detection and exclusion schemes for robust railway navigation. In: European navigation conference (ENC), Lausanne, Switzerland, pp 36–47, 9-12 May 2017
58. Albanese A, Marradi L, Labbiento G, Venturi G (2005) The RUNE project: the integrity performances of GNSS-based railway user navigation equipment. In: Proceedings of joint rail conference, Pueblo, Colorado, 16–18 Mar 2005
59. European GNSS Agency (GSA). http://www.ersat-eav.eu/
60. ESA. https://business.esa.int/projects/sbsrails
61. Rispoli F, Filip A, Castorina A, Di Mambro G, Neri A, Senesi F (2013) Recent progress in application of GNSS and advanced communications for railway signaling. In: 23rd international conference Radioelektronika, 16–17 Apr 2013
62. Neri A, Filip A, Rispoli F, Vegni AM (2012) An analytical evaluation for hazardous failure rate in a satellite-based train positioning system with reference to the ERTMS train control systems. In: Proceedings of the 25th international technical meeting of the satellite division of the institute of navigation (ION GNSS), Nashville, TN, pp. 2770–2784
63. Filip A, Baant L, Mocek H (2010) The experimental evaluation of the EGNOS safety-of-life services for railway signalling. WIT Trans Built Environ 114:549–560
64. Shin K-H, Shin D, Joung E-J, Kim Y-G (2008) The reliability and safety enhancement method of GNSS for train control application. In: The 23rd international technical conference on circuits/systems computers and Communications (ITC-CSCC)

Part II
The Railway Application

Chapter 6
The Rail Environment: A Challenge for GNSS

Salvatore Sabina, Fabio Poli and Nazelie Kassabian

Abstract This chapter describes the foundations and principles of railway signalling systems and their main key elements and provides an accurate description of the main European Rail Traffic Management System (ERTMS) properties that can be affected by the introduction of the GNSS technology. In order to bring the readers to understand the complexity of the railway environment and the main differences with respect to civil aviation and maritime environments, Sects. 6.3.1 and 6.3.2 provide an overview of the applicable European Commission Regulations and of the complete Control-Command and Signalling System suitable for obtaining the Single European Railway Area, Sects. 6.3.3 and 6.3.4 outline the reference ERTMS System Architecture with emphasis on the interfaces and the functions to guarantee the interoperability requirements, and Sects. 6.3.5 and 6.3.6 accurately describe the ERTMS dependability requirements such as safety, reliability and availability along with the related reference Mission Profile. Furthermore, Sect. 6.4 describes the current process for assessing the conformity of a single ERTMS constituent and for verifying the ERTMS Command and Control and Signalling Subsystems. Finally, Sect. 6.6 provides a quick description of the on-board train environment with respect to radio frequency interferences, multipath and non-line-of-sight conditions to outline how these phenomena, considered negligible in the civil aviation environment, have a critical role in the railway environment.

Note: The methodologies, the algorithms and the integrity concept to cope with Safety-of-Life applications described in the first part of the book will be tailored

S. Sabina (✉)
Ansaldo STS S.p.A, Via Paolo Mantovani 3-5, 16151 Genova, Italy
e-mail: salvatore.sabina@ansaldo-sts.com

F. Poli
Ansaldo STS S.p.A, Via Ferrante Imparato 184, 80147 Napoli, Italy
e-mail: fabio.poli@ansaldo-sts.com

N. Kassabian
Ansaldo STS S.p.A, Via Volvera 50, 10045 Piossasco, Torino, Italy
e-mail: nazelie.kassabian@ansaldo-sts.com

© Springer International Publishing AG, part of Springer Nature 2018

L. Lo Presti and S. Sabina (eds.), *GNSS for Rail Transportation*, PoliTO Springer Series, https://doi.org/10.1007/978-3-319-79084-8_6

for the rail application throughout this chapter and the following one. This chapter brings the reader to understand the complexity of the rail environment and the basic current principles for determining the safe train position. Refer to the glossary at the end of this book for a quick access to the definition of some railway terminology [1] that is extensively used in the explanation of these principles. Instead, the next chapter will describe the concept of Virtual Balise, proposed for the introduction of the GNSS technology, and the possible enhanced ERTMS architecture suitable for a safe implementation of this concept. Many of the concepts and techniques described in this chapter and in the first part of the book will be tailored based on the Virtual Balise concept. The enhancement of the ERTMS standard for introducing the GNSS technology shall have to guarantee both the backward compatibility with existing ERTMS solutions and the interoperability requirements; however, as the definition of this standard evolution is still in progress, the tailored solution described in Chap. 7 must be considered as a consolidated and mature working framework only. This framework is based on the experience gained in many important R&D projects such as ESA-sponsored 3InSat [2] and SBS [3] projects, and GSA-sponsored ERSAT-EAV [4], RHINOS [5] and STARS [6] projects.

6.1 Foundations of a Signalling System

The main objective of railway signalling systems is to enable safe train movements. As a train run on a railway track, a railway signalling system must appropriately rout trains on the railway tracks and space them so as to avoid collisions with one another. Furthermore, due to the high kinematic energy associated with a moving train, the detection of events by the driver that require the need to slow down and stop ahead does not guarantee the required safety, e.g. the presence of (*a*) fog that reduces the driver visibility, (*b*) another train just stopped after a curve and (*c*) a train stopped inside a gallery. Therefore, at least, a signalling system must send warnings to the driver to recommend actions (modern signalling systems do more than this, as described in the next ERTMS sections). The basic principle foundation of many signalling systems is signals and block systems. A signal is a medium to convey a particular predetermined meaning in non-verbal form; many different types of signals have populated the railway signalling systems, from the old semaphore signals to the LED Multiple Aspect Colour Light Signals. On the other end, with regard to the concept of railway block systems, each line is divided into block sections, and normally only one train is permitted to be in each block section at any one time (there can be some exceptions, but they require the application of special safe procedures). A signal is normally installed at the start and the end of each block section to enable the train to enter into and to exit from the block. The length of the block sections depends on the national signalling rules; for example, the length of Italian Block Section is 1350 or 1200 m in accordance with the type of block system. With the years, these elementary signalling systems have evolved to provide the following key functions:

- Safety Functions:
  - To prevent trains being routed in conflicting routes;
  - To maintain a safe separation distance between trains;
  - To protect trains from driver malfunction (incapacity/inattention/misjudgement);
  - To protect trains from faults of trackside and/or on-board components;
  - To ensure trains do not exceed their permitted speed, dynamically computed on-board.
- Non-Safety Functions:
  - To maximize the line capacity;
  - To automatically route trains and regulate their flow;
  - To collect diagnostic data for managing defects and providing predicative maintenance;
  - To manage the distribution of information to the public at railway station and to the on-board passengers.

6.2 Main Key Elements of a Signalling System

The following section highlights the key elements of a signalling system. Those key elements that are only ERTMS elements are described in Sect. 6.3.

6.2.1 Train Detection Unit

The train detection is a trackside function that aims to determine if a particular section (block) of track is occupied by a train or a bogie. Almost all the train detection units automatically perform such a detection by using track circuits or axle counters.

Track Circuits—They use insulated sections of the rails as an electrical circuit, which the wheels of a train or a bogie shunt when it enters the section. Conceptually, a track circuit can be represented as depicted in Fig. 6.1, where in the simplest form, the transmitter is a voltage generator and the detector is an electro-mechanical relay.

A low-voltage current is injected in the direction opposite to train movement by a transmitter; at the other end of circuit, the receiver (a relay, in older systems) keeps the signal at proceed aspect, see Fig. 6.1. When the leading axle of train or bogie enters the circuit, it establishes a low-resistance connection (short circuit) between the rails. No more current reaches the receiver, which causes the signal to show stop aspect, see Fig. 6.2.

The status of each track circuit is periodically acquired via a wired communication and processed by the interlocking (i.e. a safe platform that implements core signalling functions, see Sect. 6.2.4) or by safe equipment installed in specific location along

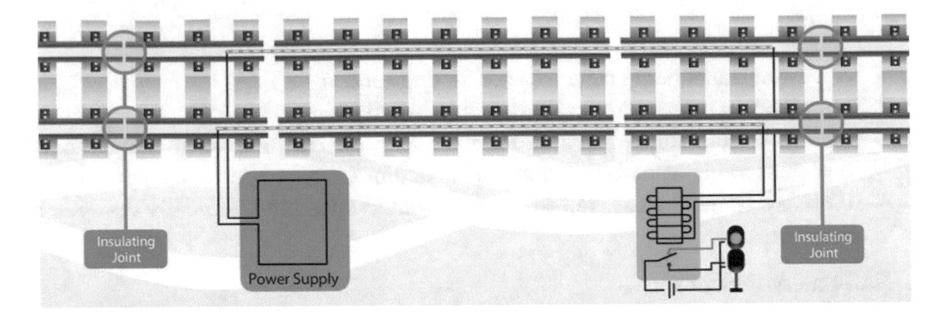

Fig. 6.1 Free track circuit

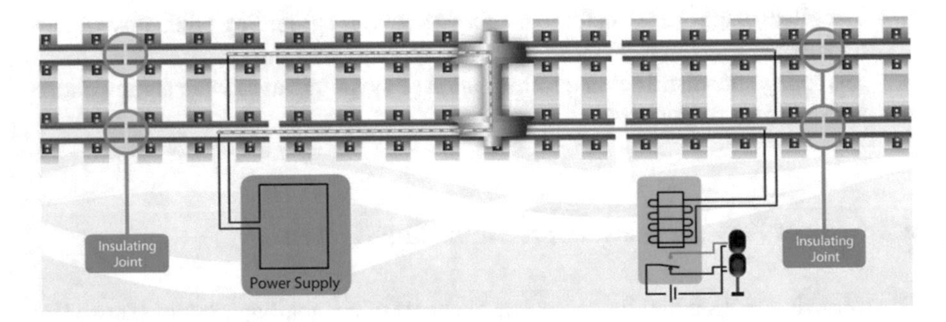

Fig. 6.2 Occupied track circuit

the railway line. The interlocking or the safe equipment commands the status of the signals to the rear of the train and to the front on single track to the danger aspect.

Many much more sophisticated types exist using (*a*) frequency signals to repeat the status of signals to on-board, (*b*) coded audio frequency signals and (*c*) frequency-shift keying (FSK).

The last two types were developed to provide immunity from EMI generated by electric trains and to also replace mechanical insulated joints with electrical joints (the use of "jointless" track circuits avoids the cut of rails and the use of insulating joints), see Fig. 6.3. The frequency of track circuits depends on electric traction type: in DC traction, 50 Hz or higher frequency is used, whereas, in AC traction systems, high-frequency circuits are used. For example, the section of an audio frequency track circuit can have a length up to 2000 m.

In addition to the train detection function, a coded track circuit also continuously transmits information to a train because it essentially acts an inductive system that uses the rails as transmission line. The coded track circuit systems eliminate the need for specialized beacons (balises).

The track circuit also illustrates the key principle of "Fail Safe" applied to all traditional signalling equipment: any break in the circuit between the transmitter and the receiver has the same functional effect as a train shunting the rails, and hence, the system fails to a safe state.

Fig. 6.3 Jointless track circuit

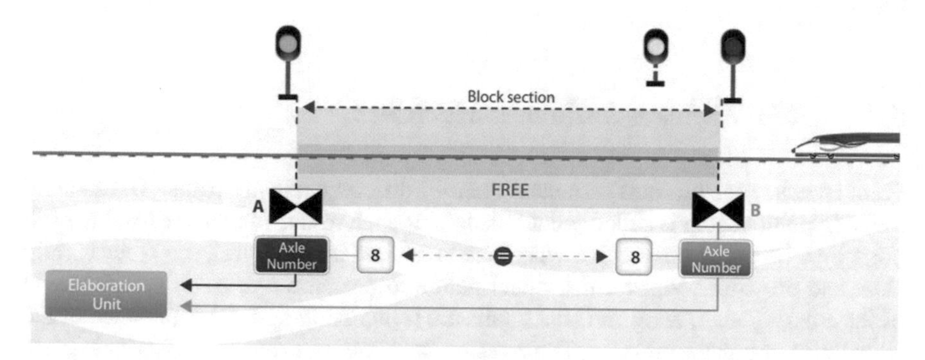

Fig. 6.4 Axle counter

By and large, in multiple aspect signalling installations, large numbers of track circuits cover the entire track layout to provide complete train detection. In a conventional automatic block system, permissible headway between trains is determined by the fixed length of each block system and is therefore invariable. Modern electronics has made possible a so-called moving block system, in which block length is determined not by fixed track distance but by the relative speeds and distance from each other of successive trains.

Axle Counters—They operate simply by counting the axles of a train entering and leaving a section of track. If the section is initially clear, then any net number of axles in the section implies that the track section is occupied, see Fig. 6.4.

Early axle counters used a mechanical lever connected to a relay; wheel flange moves lever, and each movement is counted by relay on/off. This system was suitable only for low speed.

Modern axle counter detectors have a couple of coils on the two sides of each rail; metal wheel causes a variation in magnetic flux induced by a coil and detected by the other one. Advantages of axle counter with respect to track circuits are represented by simple maintenance and no need to install equipment along the line.

However, axle counters may "forget" how many axles are in a section for various reasons such as a power failure. A manual override is therefore necessary to reset the system. This manual override introduces the human element which may be unreliable. Axle counters only provide intermittent positive indication of a rail vehicle as it passes a fixed location. If the counter unit fails or becomes disconnected, a train will pass undetected into a block that would otherwise be regarded as unoccupied. An additional limitation of axle counters is the difficulty to maintain correct counts when train wheels stop directly on the counter mechanism; this is known as "wheel rock". This can prove problematic at stations or other areas where coaches/wagons are shunted, joined and divided.

Finally, magnetic shoe brakes may also interfere with detector operation. From an operation point of view, axle counter sections are long, and this limits the line capacity.

6.2.2 Point Switches and Point Machines

Point switches (or turnouts) are used to build different railway paths. The moving part of a point switch is called point "blade" for each route. Blades are fixed to each other by a tie bar to ensure that when one is against its stock rail, the other is fully clear and provides room for the wheel flange to pass through cleanly. Either side of the crossing area, wing and check rails are provided to assist the guidance of the wheelsets through the crossing. The crossing (or "frog") can be cast or fabricated; it is made of hardened steel to increase resistance to wear. Figure 6.5 shows the schematic of a point switch.

To enable high-speed operations on point switches, long switches with moving frogs are used to reduce the impact on the wheels. Several layouts of switches are possible, when the space is limited and more routes are requested.

A point machine is the machine for remotely moving turnout blades; a point machine normally includes motor and the set of contacts required to command motor and confirm the points are moved and locked in the correct position for the route set. Motor may be placed to one side of the track or, when reduced space is available, in a special sleeper between the rails.

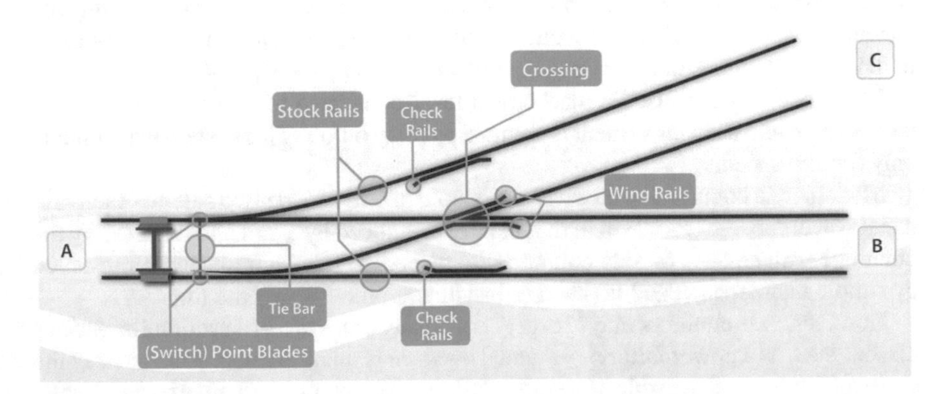

Fig. 6.5 Point switch

When long and heavy blades must be moved, e.g. for high-speed point machines, these point machines use hydraulic cylinders instead of electric motors.

6.2.3 Signals

Signals are used to convey particular predetermined meaning in non-verbal form to train drivers. Signals fall into two key types, semaphore and colour light. Semaphore signals are very old signals and, then, are used in old railway lines. Instead, colour light signals are the typically used; there are many further subtypes, conveying particular messages to the driver. Colour light signals are evolving from incandescent bulbs and lens systems to arrays of light emitting diodes (LEDs). The meaning of a signal and its format normally depend on national rules. However, a common classification is the following:

- Main signals: they mark stopping locations and cannot be passed at stop without specific authorization. Typical main signals are signals in stations.
- Permissive signals: they mark stopping locations, but may be passed at stop according to operational rules in use. Block signals (signals along the plain line) are a typical example of permissive signals.
- Distant signals: they do not mark a stopping location, and they are used to warn the driver about the aspect of next signal. Typical use of distant signals is before station entry signal to inform the driver if entry route has been set.

6.2.4 Interlocking

The underpinning functions of the interlocking can be summarized as follows:

- To maintain the current record of the position of every train in the control area and of the status of all signalling outputs (position of points/aspects displayed by signals, track circuits, axle counters, etc.); the controlled area can cover a single station or a multistation area;
- To process the signalling manager's input requests to set a particular route or to swing a set of points, and to determine if these requests lead to a safe scenario given the current recorded situation. If a safe scenario is guaranteed, then, the interlocking coherently commands and controls signalling outputs.

There are mainly two types of interlocking: relay- or computer-based interlocking.

Relay-Based Interlocking—These were mainly developed until 2000s. They carry out the interlocking functions by means of large numbers of electro-mechanical safe relays and hard-wired among them. Each specific location normally requires specific wire connections.

Computer-Based Interlocking—Instead of using safe relays and complex/hard-wired connections, they use a safe microprocessor platform where the interlocking functions are executed and microprocessor-based safe field controllers as input devices and actuators. The safe interlocking platform has normally a 2oo3 or redundant 1oo2 safe architecture. The interlocking communicates with individual field controllers by means of a duplicated serial data link for redundancy purpose and communication protocols compliant with the required safety and security railway standards. The field controllers can interface parallel inputs and outputs at an appropriate power level for individual signal and switching points. The required functionality at a particular location is implemented via a logic engine and logic data/configuration data.

6.2.5 Balises

A balise (beacon) is a physical equipment installed on a sleeper (e.g. wood or concrete sleeper). The balise does not require external power supply; it is activated/energized by a specific equipment and related antenna installed on a train.

The function of a balise is mainly to send information to the on-board that energizes/activates it. The information to be sent can be a fixed pre-stored information (i.e. this balise is named fixed balise) or a variety of information based on the input information the balise receives (i.e. this balise is named switchable balise). There are many different types of balises depending on the signalling system they belong to, e.g. KVB balise (Fig. 6.6), ASFA balise (Fig. 6.7), EBICAB balise (Fig. 6.8), ERTMS balise (Fig. 6.9). However, independently of the type of balise, the on-board subsystem is equipped with an antenna/a sensor to read the information a balise sends.

Apart from specific cases (e.g. temporary speed restrictions), the location of a balise is selected during the design phase of the signalling system and depends on the specific signalling rules to be applied. For example, it is located at a point where a change of the speed limit must be communicated to the on-board equipment or at point close to a signal to repeat the status of the signal (e.g. red) to the on-board equipment (i.e. the on-board automatically stops the train when reads this information). The number of information bits a balise can send to the on-board has increased during the evolution of the railway signalling systems, starting from 1 bit of information to 1023 bits of information in each ERTMS balise. Due to this large number of information a balise can send and the well-consolidated robust balise design, it is used in many signalling systems (not only ERTMS) such as, for example, Chinese Train Control System versions CTCS-2 and CTCS-3, SCMT in Italy, TBL1+ in Belgium and GNT in Germany. The different types of information that an ERTMS balise can send to on-board are described in [7], and those related to the train localization function are also described in Sect. 6.3.4.

Fig. 6.6 KVB balise

Fig. 6.7 ASFA balise

Fig. 6.8 EBICAB balise

Fig. 6.9 ERTMS balise

6.2.6 On-Board Unit

It is the part (software and/or hardware) of the on-board equipment which implements the on-board functions of the railway signalling system with the aim of supervising vehicle operations.

Many different signalling systems have been developed. The simplest systems logically "repeat" the trackside signal aspect received from the balise or the coded track circuit and activate an audible warning to sound in the train cab (driving position); if the train driver fails to respond appropriately, after a short interval the train brakes are automatically applied. More sophisticated systems display the maximum permitted speed and dynamic information for the route ahead, based on the distance in front which is clear and the braking characteristics of the train. In modern signalling systems, permission for a train to run to a specific location within the constraints of the infrastructure is provided through a special information known as Movement Authority (MA).

Depending on the signalling system, in general, the on-board unit is composed of:

- A safe platform where the signalling functions are executed;
- A train interface unit that provides the interface between the safe platform and the train;
- Modules and related antennas/sensors for (*a*) receiving intermittent (i.e. via balises) and/or continuous (e.g. via coded track circuits) transmissions between trackside and train and (*b*) for transferring application information to the safe platform;
- A driver machine interface to enable direct communication between the safe platform and the driver. In old and basic signalling systems (e.g. Train Stop Signalling Systems), this interface was missing.

The on-board unit is also responsible for guaranteeing an acceptably safe communication channel with the trackside for implementing the intermittent and/or continuous transmissions from trackside to train; this communication channel can be based on only a dedicated radio frequency channel (e.g. ASK modulation [8] or FSK modulation [9] with specific channel codes) or a more sophisticated channel based on a radio network (e.g. Global System for Mobile communications-Railways (GSM-R)) and a Protocol Safety Layer such as Euroradio [10] or a combination of both.

Furthermore, the existing current on-board units use one or more braking curves to continually supervise the train movement, where a braking curve is a prediction of the train speed decrease versus distance done by means of a mathematical model of the train braking dynamics and of the track characteristics ahead. This supervision is based on (*a*) the on-board estimated train front-end position and the on-board estimated train speed, computed in accordance with the physical characteristics of the train and the odometer working conditions and expressed as a distance from a location reference detected by the on-board, and (*b*) the train position confidence interval that is the distance interval within which the on-board unit assumes the actual train position is, with a defined probability and including odometer over-reading and under-reading errors.

The on-board estimated train front-end position and the on-board estimated train speed are computed with a complex odometry algorithm from the data measured by normally angular speed sensors positioned on independent wheels or a combination of more sophisticated multikinematics sensors. As a simple example, with regard to a solution based on wheel sensors, when the adhesions between the wheel and rail are good, pure rolling conditions occur and train speed can be calculated simply as follows:

$$v = R_k \times \omega_k \tag{6.1}$$

where v is train speed, ω_k is the kth wheel angular speed and R_k is the kth wheel radius. Train position is obtained simply by integrating the estimated speed.

On the other hand, when adhesion conditions between the wheel and the rail are poor, the above equation no longer holds and

$$v - R_k \times \omega_k = \delta_{vk} \tag{6.2}$$

where the so-called sliding δ_{vk} is greater than zero during a braking phase and is lower than zero during a traction phase. A good odometry algorithm compensates the difference between the wheel translation velocity and the train speed when poor adhesion conditions between the rail and the wheels occur. In order to reset the accumulated odometry error during the train travelled distance due to both the integration process of the estimated speed and abnormal slip and slide conditions, the railway signalling system foresees specific mechanisms such as the use of linked balises as explained in Sect. 6.3.4.2.

6.3 The ERTMS Standard

6.3.1 The European Commission Directives and Regulations

Directive 2008/57/EC of the European Parliament and of the Council of 17 June 2008 [11] defines a rail system as made up of the following subsystems, either:

- structural areas:
  - infrastructure: the track, switch points, engineering structures (bridges, tunnels, etc.), associated station infrastructure (platforms, zones of access, including the needs of persons with reduced mobility, etc.), safety and protective equipment;
  - energy: the electrification system, including overhead lines and on-board parts of the electric consumptions measuring equipment;
  - Control-Command and Signalling: all the equipment necessary to ensure safety and to command and control movements of trains authorized to travel on the network;

  - rolling stock: structure, command and control system for all train equipment, current-collection devices, traction and energy conversion units, braking, coupling and running gear (bogies, axles, etc.) and suspension, doors, man/machine interfaces (driver, on-board staff and passengers, including the needs of persons with reduced mobility), passive or active safety devices and requisites for the health of passengers and on-board staff.

- or functional areas:

  - traffic operation and management:
    the procedures and related equipment enabling a coherent operation of the different structural subsystems, both during normal and degraded operation, including in particular training and train driving, traffic planning and management;
    the professional qualifications which may be required for carrying out cross-border services;
  - maintenance: the procedures, associated equipment, logistics centres for maintenance work and reserves allowing the mandatory corrective and preventive maintenance to ensure the interoperability of the rail system and guarantee the performance required;
  - telematics applications for passenger and freight services:
    applications for passenger services, including systems providing passengers with information before and during the journey, reservation and payment systems, luggage management and management of connections between trains and with other modes of transport;
    applications for freight services, including information systems (real-time monitoring of freight and trains), marshalling and allocation systems, reservation, payment and invoicing systems, management of connections with other modes of transport and production of electronic accompanying documents.

The Control-Command and Signalling subsystem, also known as railway signalling subsystem, is the safety part of the rail system; it safely controls the movements of trains; it is responsible for setting up non-conflicting and safe routes for any trains authorized to travel on the controlled network.

Many different railway signalling subsystems have been developed and almost every single country used to have its own Automatic Train Protection (ATP), see Fig. 6.10 for the distribution of national ATPs in Europe [12]. These ATP systems are normally not compatible with each other; thus, in order to cross a signalling area border, the leading engine is equipped with the ATP of the leaving area and the ATP of the entering area (each of which must be self-functioning). The train drivers is normally replaced at the area border because a train driver has the driver's licence for its own country and the driver is required to apply specific manual procedures for switching off and power on the related ATPs.

The need of increasing international railway services with even better performance, the market requests of reducing the rail service costs and the objective of improving and harmonizing the level of safety among the European countries led

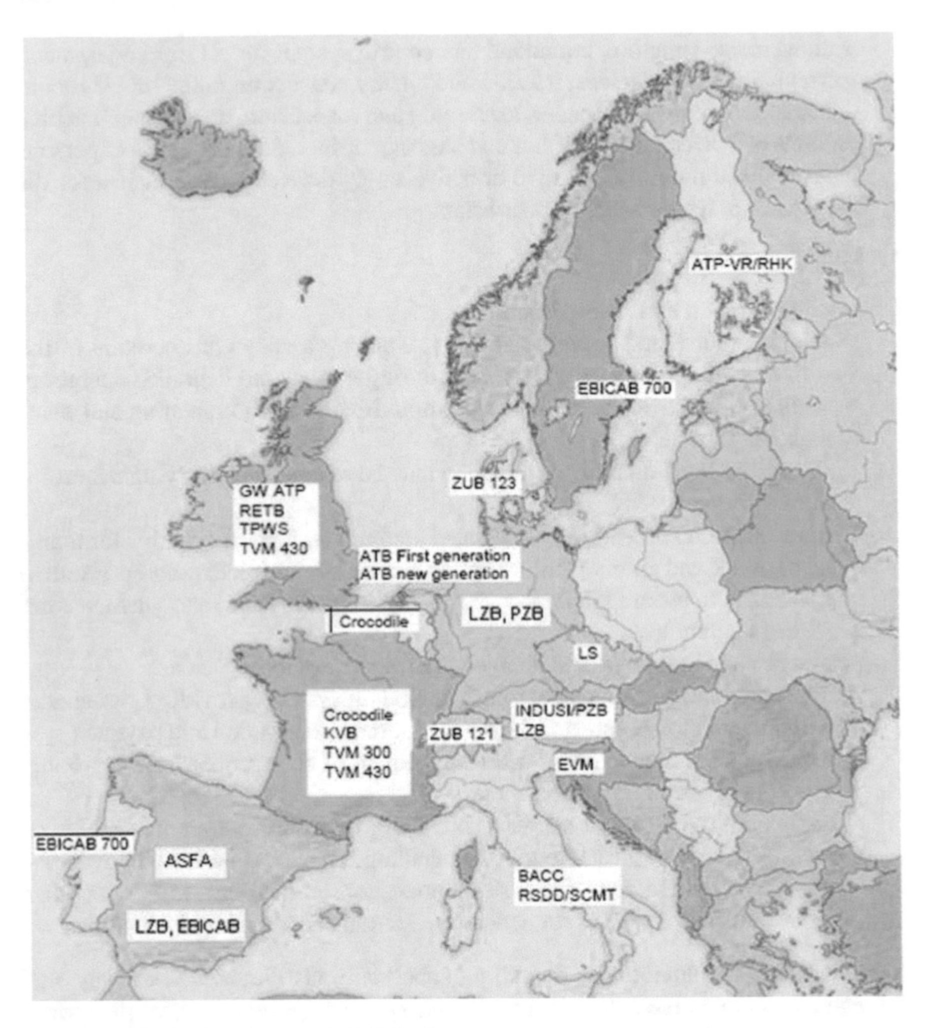

Fig. 6.10 Distribution of national ATPs in Europe

to the initiative of creating an interoperable standard for the European ATP; this initiative was launched by the European Commission in 1989 with the objective of defining the European Rail Traffic Management System (ERTMS). The fundamental objective of ERTMS is the development and the deployment of **a single harmonized control, command, signalling and communication system** that is fully interoperable across borders that can be provided from many suppliers and whose evolution is based on compatibility. The main expected advantages associated with the use of ERTMS can be summarized as:

- Cross-border interoperability;
- Improvement of the safety of national and international train traffic;

- Improvement of international passengers and freight train traffic management;
- Shorter headway on heavily trafficked lines, by driving on moving block, enabling exploitation of maximum track capacity;
- The possibility of an incremental and modular introduction of the new technology;
- Enabling Pan-European competition between the manufacturers of ERTMS/ European Train Control System (ETCS) components. Strengthening the position of the European railway industry on the world market;
- Enabling preconditions for future harmonization in other areas of rail traffic management.

In 1985, the ERTMS Users Group, named EEIG, composed of main Infrastructure Managers (IMs) and Railway Undertakings (RUs) was formed [13]. Later on, in 1998/1999, an industrial consortium, named UNISIG and made up of the main railway industries, was created. The main objectives of ERTMS EEIG and UNISIG were the definition of the first version of the ERTMS functional specifications; such specifications define ERTMS as the backbone for a digital railway system composed of the European Train Control System (ETCS) and the GSM-R.

During the last decades, many organizations, consortium and groups have been setup aimed at

(*a*) The definition and the corrective functional maintenance of the ERTMS specifications;
(*b*) The development and the delivery of the ERTMS products compliant with the ERTMS specifications;
(*c*) Supporting a harmonized ERTMS implementation;
(*d*) Providing the EU Member States and the Commission with technical assistance in the fields of railway safety and interoperability.

Interoperability and compatibility are collective European issues that, for their overcoming, requires the cooperation between Member States, IMs and RUs. To this end, Annex II of Directive 2008/57/EC defined the Control-Command and Signalling (CCS) Subsystems of the railway system in the European Union and the same Directive also provided the Technical Specification for Interoperability (TSI) document [11] relating to these CCS subsystems.

Then, Commission Regulation (EU) 2016/919 of 27 May 2016 [14] confirmed that the Control-Command and Signalling Subsystems are defined as "all the equipment required to ensure safety and to command and control movements of trains authorized to travel on the network". Note that "network" means all or part of any Member State's railway network on the day this Regulation enters into force, except when the subsystem(s) is (are) subject to renewal or upgrading.
The features of these Control-Command and Signalling Subsystems [14] are:

(*1*) The functions that are essential for the safe control of railway traffic and that are essential for its operation, including those required for degraded modes;
(*2*) The interfaces;
(*3*) The level of performance required to meet the essential requirements.

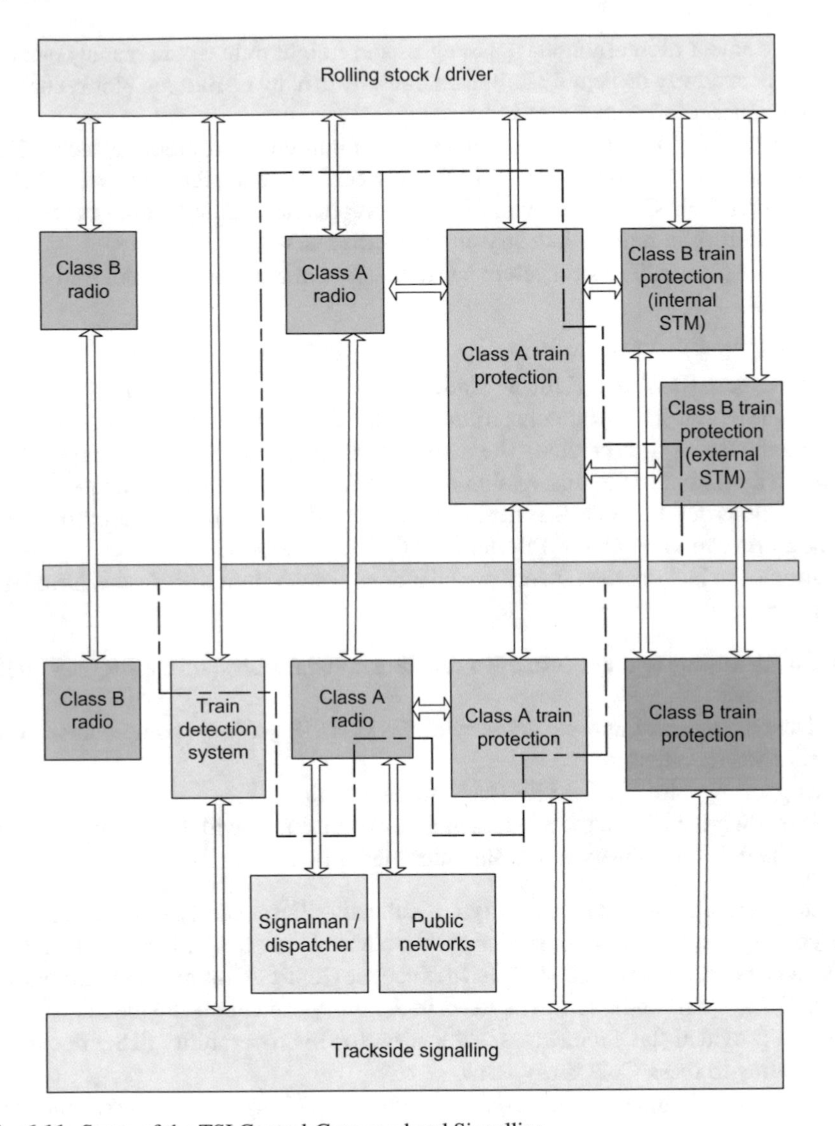

Fig. 6.11 Scope of the TSI Control-Command and Signalling

Figure 6.11 gives an overview of the scope of the TSI Control-Command and Signalling [15]. The TSI specifies only those requirements which are necessary to ensure the interoperability of the Union railway system and the compliance with the essential requirements.

The Control-Command and Signalling Subsystems, as defined in the TSI, include the following parts:

(*1*) Train protection;
(*2*) Voice radio communication;
(*3*) Data radio communication;
(*4*) Train detection.

The Class A train protection system is ETCS while the Class A radio system is GSM-R. The requirements for the Control-Command and Signalling On-board Subsystem are specified in relation to Class A radio mobiles and train protection, whereas the requirements for the Control-Command and Signalling Trackside Subsystem are specified in relation to:

(*1*) The Class A radio network;
(*2*) The Class A train protection;
(*3*) The interface requirements for train detection systems, to ensure their compatibility with rolling stock.

For Class A train detection, the TSI specifies only the requirements for the interface with other subsystems. Class B systems for the trans-European railway system network[1] are a limited set of train protection legacy systems that were in use in the trans-European railway network before 20 April 2001. Class B systems for other parts of the rail network (i.e. it comprise the high-speed rail network and the conventional rail network) of the rail system in the European Union are a limited set of train protection legacy systems that were in use in that network before 1 July 2015. The list of Class B systems is reported in the European Railway Agency technical document [12].

The Commission Regulation [14] requires that Member States ensure that the functionality, performance and interfaces of the Class B systems remain as currently specified, except where modifications are needed to mitigate safety-related flaws in those systems. Class B systems significantly hamper the interoperability of locomotives and traction units but are necessary to ensure safe operations where Class A systems are not implemented. Therefore, it is required to avoid creating additional obstacles to interoperability by, for example, altering these Class B systems or by introducing new systems.

The European deployment plan [17] of ERTMS aims to ensure that vehicles equipped with ERTMS can gradually have access to an increasing number of lines, ports, terminals and marshalling yards without needing Class B systems in addition to ERTMS. A deployment plan for the core network corridors must also include stations, junctions, access to core maritime ports and inland ports, airports, rail/road terminals and infrastructure components as they are essential to achieve interoperability in the European railway network.

[1]The trans-European network comprises transport infrastructure, traffic management systems and positioning and navigation systems. The transport infrastructure includes road, rail and inland waterway networks, motorways of the sea, seaports and inland waterway ports, airports and other interconnection points between modal networks. A complete description is provided in [16].

The Commission Regulation [14] also requires a transparent plan for European Train Control System (ETCS) implementation and for the decommissioning of Class B systems to achieve the objectives of the Single European Railway Area.

The TSI does not apply to existing "trackside Control-Command and Signalling" and "on-board Control-Command and Signalling" subsystems of the rail system already placed in service on all or part of any Member State's railway network on the day the Regulation enters into force, except when the subsystem is subject to renewal or upgrading.

As far as ETCS is concerned, many different baselines were delivered during the last years; for example, ETCS Baseline 2 was enforced in 2008 and the second release of ETCS Baseline 3 was delivered in 2016. Baseline 3 is fully backward-compatible as demonstrated in the report "baseline compatibility assessment", published by ERA, and it is considered a mature and functionally complete set of specifications to be maintained stable for a long period, protecting investment in on-board and trackside implementations and for enabling widespread deployment, see the ERA report [18].

Table 6.1 reports:

- The main organizations, consortium and groups that have an important role in the definition, maintenance and deployment of ERTMS;
- A short description of their missions;
- Their membership.

6.3.2 ERTMS/ETCS System Structure

Due to the nature of the signalling functions, the ERTMS/ETCS system is partly implemented on the trackside and partly on-board the trains. Thus, it is composed of two subsystems: the on-board subsystem and the trackside subsystem. On the other side, the ERTMS/ETCS environment is made up of:

- The train, which is considered in the train interface specification;
- The driver, which is considered via the driver interface specification;
- Other on-board interfaces such as, for example, on-board recording interface, radio interface, Eurobalise interface;
- External trackside systems (interlockings, control centres, etc.), for which no interoperability requirement is established.

Depending on the ERTMS/ETCS application-level Subset 026-2 [7], which is defined according to the functions and the system performance required, the trackside subsystem can be composed of different types of trackside components. Only some of these trackside components are reported herein:

- Balise—It is a transmission equipment installed on the track sleepers that can send information, named telegrams, to the on-board subsystem. Balises can be

Table 6.1 Main organizations, groups, consortia related to ERTMS

Name	Mission	Organization
European Union Agency for Railways (ERA)	The Agency has been established to provide the EU Member States and the Commission with technical assistance in the fields of railway safety and interoperability. This assistance involves the development and implementation of Technical Specifications for Interoperability and a common approach to questions concerning railway safety. The Agency's main tasks are: 1. promoting a harmonized approach to railway safety; 2. devising the technical and legal framework to enable removing technical barriers and acting as the System Authority for the Single European Train Control and Communication System; 3. improving accessibility and the use of railway system information; and 4. acting as the European Authority under the 4th Railway Package issuing vehicle authorizations and safety certificates, while improving the competitive position of the railway sector	Following the entry into force of the technical pillar of the 4th EU Railway Package on 15th June 2016, the European Union Agency for Railways replaces and succeeds the European Railway Agency. See http://www.era.europa.eu for details
ERTMS Users Group, a European Economic Interest Grouping (EEIG)	The ERTMS Users Group combines the knowledge and experience of its members to support the introduction of ERTMS by ensuring a safe, reliable and interoperable system solution at reasonable and affordable costs. The ERTMS Users Group advises the Community of European Railway and Infrastructure Companies (CER), the European Rail Infrastructure Managers (EIM), the European Rail Freight Association (ERFA), the European Passenger Train and Traction Lessors Association (EPTTOLA) and the International Union of Railways (UIC) about the ERTMS deployments	The members and cooperating railway companies are: • ADIF, Spain, • Banedanmark, Denmark, • DB , Germany, • Infrabel, Belgium, • BaneNOR, Norway (cooperating railway company), • Network Rail, Great Britain, • ProRail, the Netherlands, • SNCF Rseau, France, • RFI, Italy, • Trafikverket, Sweden, • SBB, Switzerland (cooperating railway company)

Table 6.1 (continued)

Name	Mission	Organization
UNISIG	The role of UNISIG is the development, maintenance and updating of the ERTMS specifications in close cooperation with ERA. UNISIG actively contributes, together with the railway representative bodies, to the various related working groups of the agency and to the ERTMS specification and its maintenance	The UNISIG Consortium is an Associated Member of UNIFE. Seven companies now known as Alstom, Ansaldo STS, Bombardier, Siemens, Thales, CAF and AD Praha are its Full Members. MERMEC became Associated Member in 2010
UNIFE	UNIFE's purpose is to represent its members' interests at international and EU levels. The UNIFE mission is to proactively foster an environment where its members can provide competitive railway systems for the growing demand for rail transport. UNIFE and its members also work on the setting of interoperability standards and coordinate EU-funded research projects that aim at the technical harmonization of railway systems	UNIFE is representing the European rail manufacturing industry in Brussels since 1992. The Association gathers over 80 of Europe's leading large and SME rail supply companies active in the design, manufacture, maintenance and refurbishment of rail transport systems, subsystems and related equipment. UNIFE also brings together 14 national rail industry associations of European countries. UNIFE members have the 84% market share in Europe and supply 46% of the worldwide production of rail equipment and services

Table 6.1 (continued)

Name	Mission	Organization
Railway Operational Communications Industry Group (GSM-R Industry Group in the past)	In 2000, the GSM-R Industry Group (IG) was founded as an industry organization to actively promote the GSM-R technology and ensure successful deployment of GSM-R projects across Europe. At the beginning, the main focus of the GSM-R IG was to promote expansion of the GSM-R European standard for railway radio communication worldwide, while assuring full compliance with EIRENE standards and guaranteeing an economy of scale necessary to make the technology economically viable and maintainable. In 2016, the GSM-R Industry Group changed the name to Railway Operational Communication (ROC) Industry Group. The members of the ROC Industry Group are all GSM-R industry suppliers who are committed to GSM-R technology, and who have been investing substantial development efforts in GSM-R standards and projects	The ROC Industrial Group is composed of: • ALSTOM, • FREQUENTIS, • Funkwerk, • ISKRATEL, • Kapsch, • LEONARDO, • NOKIA, • SIEMENS, • Triorail, • WENZEL
Community of European Railway and Infrastructure Companies (CER)	CER's role is to support an improved business and regulatory environment for European railway operators and railway infrastructure companies	CER represents more than 70 members and partners. It has a diversity of members ranging from long-established bodies to new entrants and both private and public sector organizations. CER members come from EU-28, Norway, Switzerland, Moldova, EU candidate countries (Macedonia, Montenegro, Serbia, Turkey), and Western Balkan countries. CER also has partners in Japan and Georgia

Table 6.1 (continued)

Name	Mission	Organization
European Rail Infrastructure Managers (EIM)	The role of EIM is to provide a single voice to represent its members, Infrastructure Managers (IMs), vis-a-vis to the relevant European institutions and sector stakeholders. EIM also assists members to develop their businesses through the sharing of experiences and contributing to the technical and safety activities of ERA. Its mission can be summarized as follows: • To promote the development, improvement and efficient delivery of rail infrastructure in the EU; • To make liberalization a success in the countries where it has been implemented; • To represent its members' political, technical and business interests to all relevant EU institutions; • To support the business development by providing a forum for cooperation; • To provide an environment for the leaders of IMs for sharing best practices and efficiency tools	European Rail Infrastructure Managers (EIM) was established in 2002 following the liberalization of the EU railway market. Based in Brussels, EIM is registered as an international, non-profit association under Belgian law
European Rail Freight Association (ERFA)	ERFA fully supports a competitive and innovative single European railway market offering attractive, fair and transparent market conditions for all railway companies. ERFA subscribes to the EU's White Paper goal of achieving sustainable transport, moving more freight and passengers off the roads and onto rail. ERFA is supporting ERTMS to make rail transport safer and more competitive by deploying ERTMS-fitted rolling stock wherever it is sustainable	ERFA was created in Brussels (B) in 2002. ERFA represents all those operators who want open access and fair market conditions, and sustains their role of pushing forward the development of the railway market. In 2016, ERFA represents 36 members from 16 countries. The members of ERFA represent the entire value chain of rail transportation: • Rail freight operators, • Wagon keepers, • Service providers, • Forwarders, • Passenger operators, • National rail freight associations

Table 6.1 (continued)

Name	Mission	Organization
European Passenger Train and Traction Lessors' Association (EPTTOLA)	EPTTOLA aims at being a representative body for European passenger train and traction operating lessors that: • examines those activities of European authorities which impact on the interests of its Members and communicates those activities to its Members; and • works with such European authorities so as to ensure that its Members' interests are represented in those activities. EPTTOLA is recognized by the European Commission as an official stakeholder representative body for train leasing companies	Membership of the association is open to private companies that supply passenger trains and traction equipment in countries within the European Economic Area, or in countries that are applying for membership to the European Union. In 2017, EPTTOLA consists of 6 members: • Alpha Trains, • Angel Trains Ltd, • Beacon Rail Leasing Ltd, • Macquarie European Rail, • Eversholt Rail (UK) Ltd, • Porterbrook Leasing Company Ltd.
International Union of Railways (UIC)	Promote rail transport at world level with the objective of optimally meeting current and future challenges of mobility and sustainable development. Promote interoperability, and as a Standard-Setting Organization, create new world IRSs (International Railway Solution) for railways (including common solutions with other transport modes). Develop and facilitate all forms of international cooperation among Members, facilitate the sharing of best practices (benchmarking). Support Members in their efforts to develop new business and new areas of activities.	UIC was created in 1921. It initially had 51 members from 29 countries including Japan and China, which were soon joined by the railways from the USSR, the Middle East and North Africa. The UIC Members categories are:

Table 6.1 (continued)

Name	Mission	Organization
	Propose new ways to improve technical and environmental performance of rail transport, improve competitiveness, reduce costs	• **Active Members**: companies or entities, public or private, (passenger and/or freight) Railway Undertaking ensuring traction and/or a railway Infrastructure Manager that also have a volume of railway business in excess of an amount fixed by Internal Regulation R1 approved by the General Assembly. • **Associate Members**: companies or entities, public or private (passenger and/or freight) Railway Undertaking ensuring traction and/or a railway Infrastructure Manager that do not fulfil the condition for being active members. • **Affiliate Members**: companies or entities, public or private, including institutes and associations, whose railway activities relate to urban, suburban or regional services or which conduct activities linked to the rail business

organized to logically belong to a group (named Balise Group); the combination of all telegrams sent by each balise of the Balise Group defines the message sent by the Balise Group.

- The radio communication network (GSM-R)—This network is used for the bidirectional exchange of messages between on-board subsystems and Radio Block Centre (RBC) or radio infill units.
- The Radio Block Centre (RBC)—It is a computer-based system, installed in a building of a railway station, that elaborates messages to be sent to the train on the basis of (*a*) information received from external trackside systems and (*b*) information exchanged with several on-board subsystems. The main objective of these messages is to provide movement authorities to allow the safe movement of trains on the railway infrastructure area under the responsibility of the RBC. An RBC can control a railway area of some hundreds of kilometres (e.g. the RBCs installed in the Italian high-speed lines control an average area of about 200 km); a large area requires the cooperation of different RBCs.
- Radio Infill Unit—It provides signalling information in advance as regard to the next main signal in the train running direction. It is a safe platform normally installed in a small cabinet along the track.[2]

Similar to the trackside subsystem, depending on the ERTMS/ETCS functions and the system performance required, the on-board subsystem can be composed of the ERTMS/ETCS on-board equipment and the on-board part of the GSM-R radio system. The ERTMS/ETCS on-board equipment is a safe computer-based system that supervises the movement of the train to which it belongs, on the basis of information exchanged with the trackside subsystem. On the other end, the GSM-R on-board radio system is used for the bidirectional exchange of messages between on-board subsystem and RBC or radio infill unit.

6.3.3 ERTMS/ETCS Reference Architecture

Directive 2008/57/EC [11] requires that the subsystems and the interoperability constituents including interfaces meet the essential requirements set out in Annex III to this Directive. The essential requirements for Class A systems cover the following classes of requirements:

(*1*) Safety;
(*2*) Reliability and availability;
(*3*) Health;
(*4*) Environmental protection;
(*5*) Technical compatibility.

[2]Some IMs have implemented the radio infill functions by means of Central Radio Infill safe platforms installed in large stations.

On the other end, the requirements for Class B systems are under the responsibility of the relevant Member States and these requirements are not specified in this Directive.

In accordance with Directive 2008/57/EC [11], interoperability constituents are "any elementary component, group of components, subassembly or complete assembly of equipment incorporated or intended to be incorporated into a subsystem, upon which the interoperability of the rail system depends either directly or indirectly. The concept of a constituent covers both tangible objects and intangible objects such as software". Appendix A.1 (Interoperable Constituents) summarizes the basic interoperability constituents and the already identified groups of interoperability constituents.

The TSI [14] aims to guarantee the interoperability between a Control-Command and Signalling Trackside Subsystem (compliant with the TSI) and Control-Command and Signalling On-board Subsystems (compliant with the TSI). To achieve this goal:

(*1*) functions, interfaces and performances of the Control-Command and Signalling On-board Subsystem are standardized, ensuring that every train will react in a predictable way to data received from trackside;

(*2*) for the Control-Command and Signalling Trackside Subsystem, track-to-train and train-to-track communication are fully standardized.

The TSI [14] only specifies interfaces that are necessary to achieve interoperability; for interfaces between trackside and on-board, the TSI requires that their implementation (functions, protocols, electrical and physical aspects) complies with the mandatory specifications. For other interfaces (e.g. between equipment allocated either on-board or trackside), different solutions are acceptable, provided that functional and performance requirements relevant for the achievement of interoperability are met. Figure 6.12 [7] outlines the ERTMS/ETCS system architecture and its interfaces specified in the related documents named ERTMS/ETCS subsets, e.g. [7, 9, 10]. The box entities inside the ERTMS/ETCS on-board equipment box are shown only to highlight the scope of the interfaces.

In order to allow each individual railway administration to select the appropriate ERTMS/ETCS application trackside, according to their strategies, to their trackside infrastructure and to the required performance, ERTMS/ETCS has been defined as a scalable and modular system. To this end, different ERTMS/ETCS application levels (short: levels) [1, 7] have been defined and they are a way to express the possible operating relationships between trackside and train. Level definitions are related to the trackside equipment used, to the way trackside information reaches the on-board units and to which functions are executed in the trackside and in the on-board equipment, respectively.

For example, ERTMS/ETCS Level 2 is a radio-based train control system, see Fig. 6.13 [7]; each train is equipped with ERTMS/ETCS operating on a railway line controlled by a RBC and equipped with Eurobalises as spot transmission devices mainly for location referencing; the radio communication between the RBC and

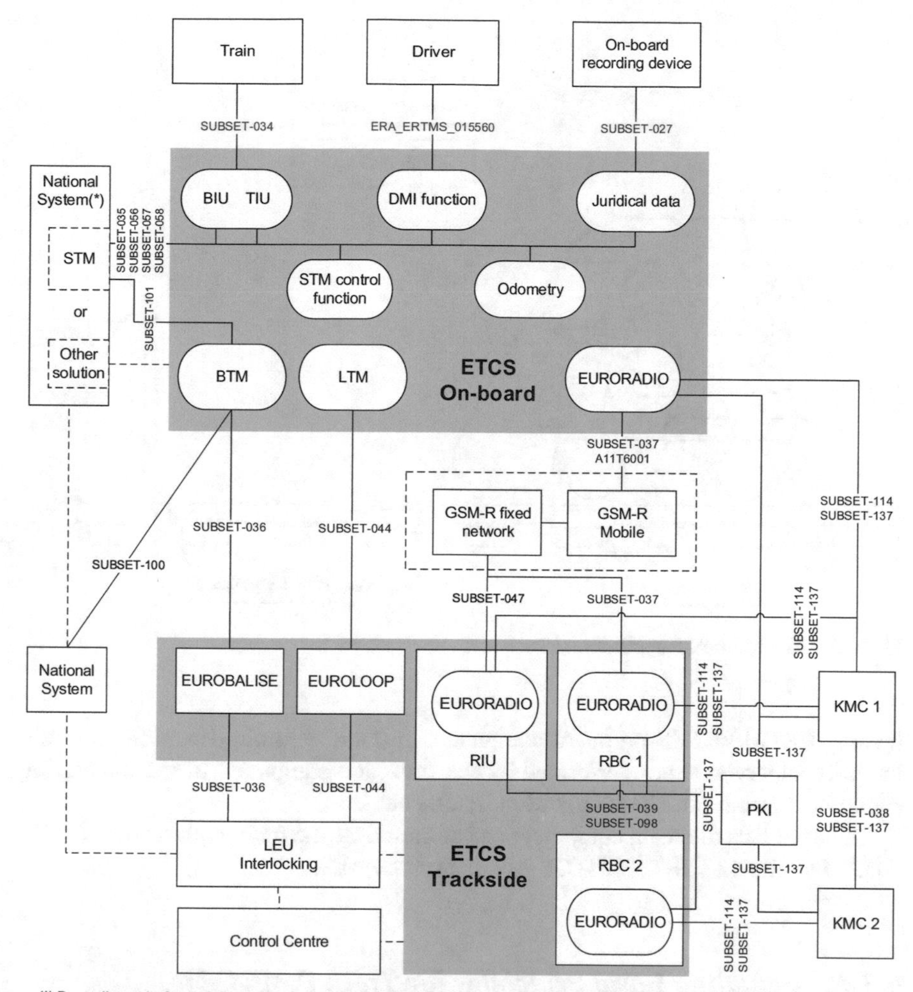

Fig. 6.12 ERTMS/ETCS system architecture and its interfaces

each equipped train is performed by using the GSM-R communication link and the Euroradio Protocol Stack [10]; this protocol stack guarantees a safe and secure radio communication via the dedicated communication session established between the RBC and the connected train.

Movement authorities defined in [1] and Subset 026-3[7] are generated by the RBC and are individually transmitted to each train; on the other side, based on the received movement authorities, the on-board ETCS provides a continuous speed supervision system, which also protects against overrun of the authority. The RBC knows each ERTMS/ETCS controlled train individually by the ERTMS/ETCS identity of its

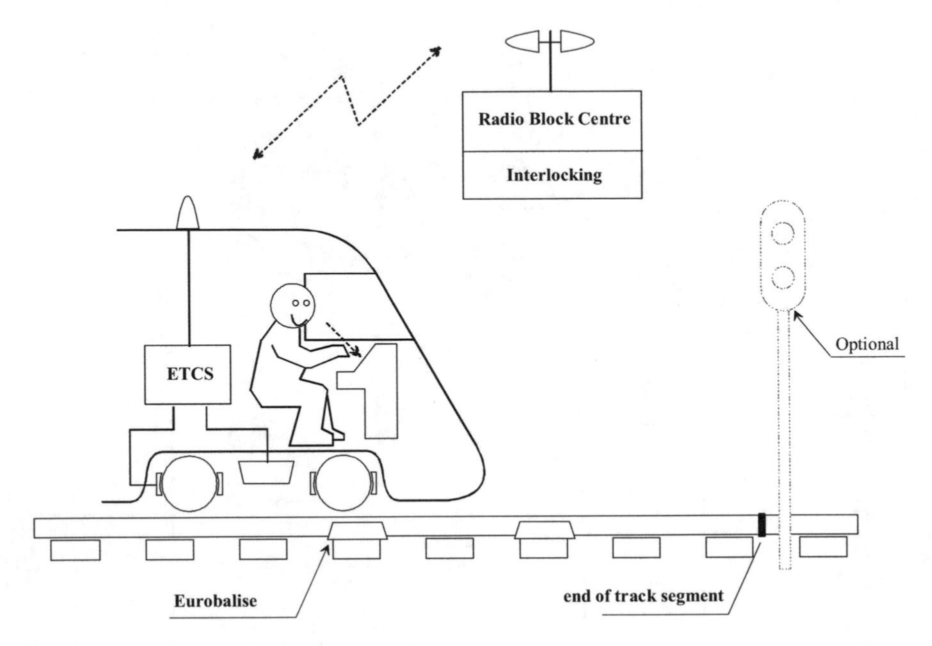

Fig. 6.13 ERTMS/ETCS application level 2

leading ERTMS/ETCS on-board equipment. In Level 2, train detection and train integrity supervision are performed by the trackside equipment of the underlying signalling system (interlocking, track circuits, etc.).

Table 6.2 lists the main equipment and summarizes the main high-level ERTMS/ETCS functions of ERTMS/ETCS Application Level 2.

6.3.4 Signalling Basic Principles for Train Positioning in ERTMS Application Level 2

This paragraph describes a **subset** of the main basic principles used in ERTMS/ETCS Level 2 for implementing the train position function [7] that might be affected from the introduction of the GNSS technology. The understanding of these basic principles is propaedeutic for the qualitative and quantitative evaluation of the enhanced ERTMS architecture, based on the Virtual Balise concept, described in the next chapter.

As Eurobalises (in short balises) are used as location references in Level 2, the description starts from the basic principles related to balises.

Table 6.2 List of the main equipment and summary of the main high-level functions of ERTMS/ETCS level 2

Trackside equipment	On-board equipment
• Radio Block Centre • Radio Telecommunication GSM-R Network • Euroradio for bidirectional track-train communication • Eurobalises mainly for location referencing	• On-board equipment with the Balise Transmission Module (module in charge of receiving telegrams from Eurobalises) and the Euroradio for managing the communication with the RBC
Main ERTMS/ETCS trackside functions	Main ERTMS/ETCS on-board functions
• Knowing, by the ERTMS/ETCS identity, each train equipped with and running under ERTMS/ETCS within a RBC area. • Following each ERTMS/ETCS controlled train's location within a RBC area. • Determining individually movement authorities for each train, according to the underlying signalling system. • Transmitting movement authorities and track description to each train individually. • Handing over of train control between different RBC's at the RBC-RBC borders	• The train reads Eurobalises and sends its position relative to the detected balises to the Radio Block Centre. • The train receives a Movement Authority and the track description via Euroradio. • Selection of the most restrictive value of the different speeds permitted at each location ahead. • Calculation of a dynamic speed profile taking into account the track description data and the train running/braking characteristics which are known on-board. • Comparison of the train speed with the permitted speed and commanding of the brake application if necessary. • Cab signalling to the driver

6.3.4.1 Balise configuration

A Balise Group can be composed of from one balise to eight balises; for example, Fig. 6.14 represents a Balise Group made up of two balises. Each balise stores at least the following information:

- The internal number (from 1 to 8) of the balise (e.g. in the example, 1 out of 2 for the first balise of the Balise Group or 2 out of 2 for the second balise);
- The number of balises inside the group (e.g. 2 in the example);
- The Balise Group identity.

The internal number of the balise describes the relative position of the balise in the Balise Group. Every Balise Group composed of at least two balises has its own one dimensional coordinate system, see Fig. 6.15. The origin of the coordinate system for each Balise Group is given by the balise number 1 (called location reference) in the Balise Group. The orientation of the coordinate system of a Balise Group (i.e. nominal or reverse direction) is identified as Balise Group orientation. The nominal direction of each Balise Group is defined by increasing internal balise numbers.

Balise Groups consisting of only one single balise are referred to as "single Balise Groups", see Fig. 6.16. A single Balise Group has no inherent coordinate system and,

Fig. 6.14 Balise Group composed of two balises

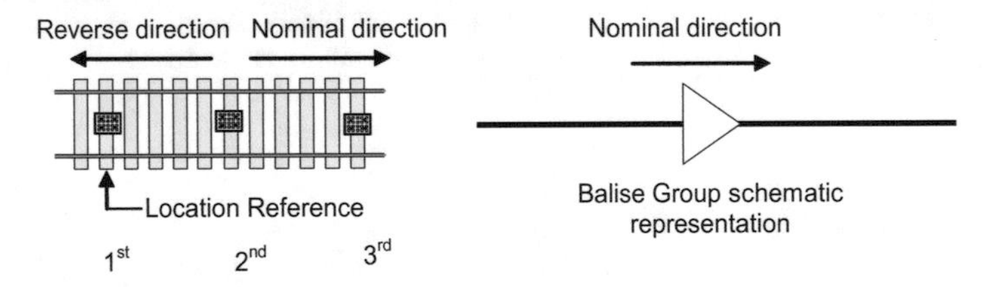

Fig. 6.15 Coordinate system and orientation of the Balise Group

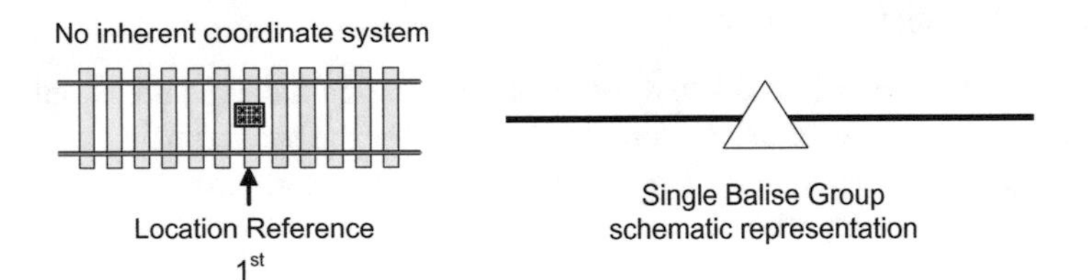

Fig. 6.16 Single Balise Group

in Level 2, RBC is responsible for dynamically assigning the coordinate system to it, see [7] for details.

A balise can contain directional information, i.e. valid either for nominal or for reverse or for both directions.

For reducing the probability of not detecting a Balise Group or of losing the information transmitted from one balise of the Balise Group, each balise can be

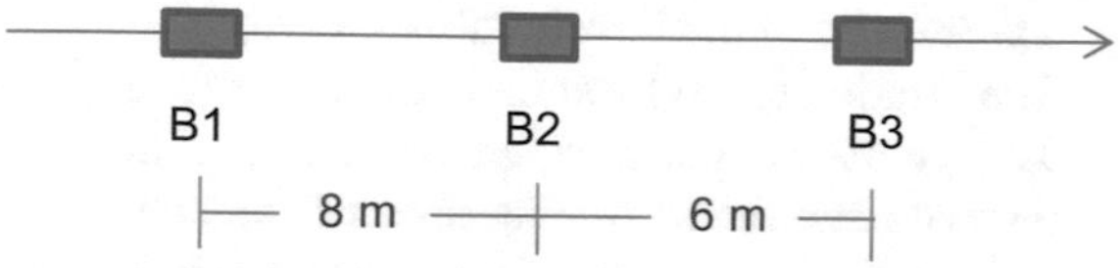

Fig. 6.17 Missed balise of a Balise Group (no balise found within 12 m)

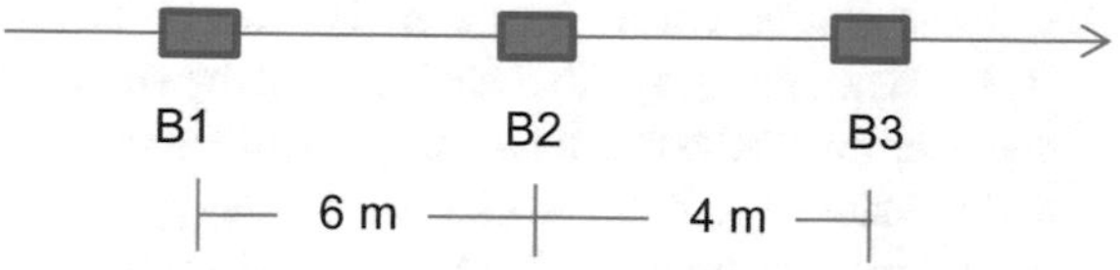

Fig. 6.18 Missed balise of a balise roup (the following balise has been passed)

duplicated (i.e. the duplicated balise contains the same signalling information of the balise to be duplicated). For example, suppose having a Balise Group composed of balise 1, balise 2 and balise 3 where balise 2 duplicates balise 1, i.e. the location reference (balise 1) of the Balise Group is duplicated by balise 2 and balise 2 is the duplicated balise of balise 1.[3] When the on-board equipment decodes the telegram of each balise, the on-board equipment understands by means of a specific value of a variable contained in the telegram if the information is the duplicated information; then, it only uses one set of duplicated information to compose the Balise Group message. Therefore, in case of the correct detection of balise 2 and balise 3 only, the Balise Group is considered correctly detected even though balise 1 has been lost. As another example, a Balise Group made up of two balises duplicating each other is treated as a single Balise Group when only one balise is correctly read. Note that the loss of a duplicated balise is an event detectable by the on-board equipment.

A balise within a Balise Group is regarded as missed if no balise is found within the maximum distance between consecutive balises from the previous balise in the group or a following balise within the group has been passed.

For example, consider a Balise Group composed of three balises BG = {B1, B2, B3}, see Figs. 6.17 and 6.18. In Fig. 6.17, B2 is considered missed if it is not detected within 12 m from the correct detection of B1.

Similarly, in Fig. 6.18, B2 is considered missed when B3 has been declared passed (i.e. detected and correctly processed) and B2 has not been detected.

6.3.4.2 Balise Linking

The detection of the loss of a Balise Group in ERTMS/ETCS can be done by using the concept of linked Balise Groups. A Balise Group is linked when the on-board ETCS knows the linking information associated with such Balise Group in advance. The linking information Subset 026−3 [7] is sent by the RBC and includes:

---

[3]The balise 1 telegram has the specific variable M_{DUP} Subset 026−8 [7] set to indicate that the information is duplicated in the balise after balise 1. On the other side, the balise 2 telegram has the specific variable M_{DUP} set to indicate that the information is duplicated in the balise before balise 2.

- The identity of the linked Balise Group;
- The location of the location reference of the Balise Group;
- The accuracy of this location reference (i.e. the absolute value in meters of the difference between the true location reference value based on the signalling design, and the actual installation position of the balise location). Note that, due to physical installation constraints, a balise can be physically installed into a different sleeper with respect to that used during the signalling design phase; the signalling designer must take this possibility into account. In addition, if the reference balise is duplicated, the trackside signalling designers are also responsible for defining the location accuracy value to cover at least the location of the two duplicated balises. The minimum and the maximum distances between consecutive balises within a group are respectively 2.3 m [9] and 12 m [19];
- The direction with which the linked Balise Group will be passed over (nominal or reverse);
- The linking reaction required if a data consistency problem[4] occurs with the expected Balise Group.

For each linked Balise Group, the trackside is responsible for commanding one of the following reactions to be used in case of data inconsistencies:

- Train trip[5];
- Command service brake[6];
- No linking reaction.

In general, the concept of linking can be used for:

- Determining whether a Balise Group (i.e. its location reference) has been missed or not found within the expectation window (e.g. a space window determined on-board in accordance with estimated accumulated odometry error, the balise detection location error of the Last Relevant Balise Group (LRBG) reference balise and the location accuracies of the LRBG and of the expected BG) and take the appropriate action;
- Assigning a coordinate system to Balise Groups consisting of a single balise;
- Correcting the train confidence interval due to odometer inaccuracy (explained later).

[4]Subset 026 provides the complete description of the data consistency checks to be executed on the information received from each balise (i.e. information named balise telegram), from the complete set of telegrams received from the set of balises in the Balise Group (i.e. complete set of telegrams named Balise Group message), from linked balises groups. Some examples of data consistency problems are named: a balise is missed inside a Balise Group, a balise is detected but no balise telegram is decoded, at least a variable inside a Balise Group message has an invalid value.

[5]In general, train trip is initiated when a train erroneously passes a specified location, named End Of Authority (EOA)/Limit Of Authority (LOA), and causes an immediate application of the emergency brake and a procedure to acknowledge the event. The emergency brake implies the application of a predefined brake force in the shortest time to stop the train with a defined level of brake performance.

[6]Service brake implies the application of an adjustable brake force to control the speed of the train, including stop and temporary immobilization.

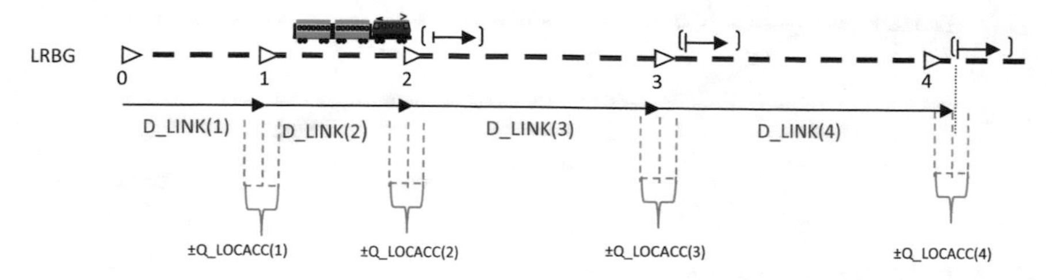

Fig. 6.19 Linked Balise Group

Figure 6.19 provides an example of linked Balise Groups where:

- The on-board ETCS has received the linking information from RBC just after the detection of the Balise Group marked in figure as LRBG (Last Relevant Balise Group).
- The received linking information states:
  - The identity of the linked Balise Group: 1, 2, 3, 4;
  - Where the location reference of each Balise Group (BG) has to be found: BG1 after D_LINK(1) space from LRBG, BG2 after D_LINK(2) space from BG1, BG3 after D_LINK(3) space from BG2 and BG4 after D_LINK(4) space from BG3;
  - The accuracy of each location reference: Q_LOCACC(1) meters for BG1, Q_LOCACC(2) meters for BG2, Q_LOCACC(3) meters for BG3, and Q_LOCACC(4) meters for BG4. For simplicity, the example assumes that there is no duplication of the location reference balises;
  - The direction with which the linked Balise Group will be passed over: nominal or reverse;
  - The reaction required if a data consistency problem occurs with the expected Balise Group: for example, no reaction.

The on-board equipment evaluates the distance information received from trackside as nominal information (without taking into account any tolerances); therefore, the balise location accuracy Q_LOCACC must be dimensioned for also including the tolerance of the linking distance. For example, Q_LOCACC(i) must also take into account the distance tolerance associated with D_LINK(i).

A Balise Group that is not marked as linked is named unlinked. An unlinked Balise Group contains information that must be processed by an on-board ETCS even when the Balise Group is not announced by linking. Unlinked Balise Groups consist at minimum of two balises. If the on-board equipment is not able to recognize whether a balise group is linked or unlinked (if none of the balises in the Balise Group can be read correctly), that Balise Group is considered as unlinked.

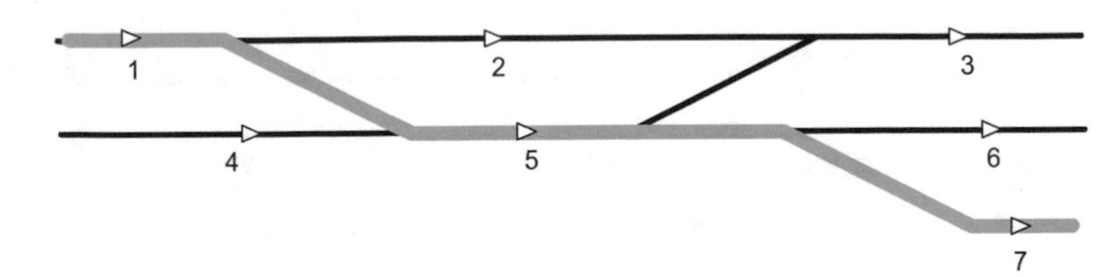

Fig. 6.20 Actual route of the train

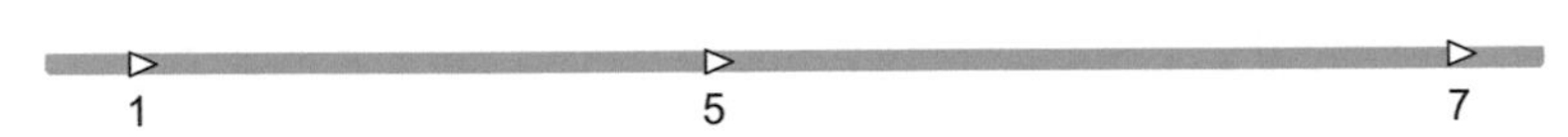

Fig. 6.21 Route known by the train

6.3.4.3 Location Principles, Train Position and Train Orientation

With regard to location principles, ERTMS/ETCS identifies two types of data:

- Data that refer only to a given location, referred to as location data (e.g. level transition orders, linking);
- Data that remain valid for a certain distance, referred to as profile data (e.g. static speed profile, gradients).

All location and profile data transmitted by a balise refer to the location reference and orientation of the Balise Group to which the balise belongs. On the other side, all location and profile data transmitted from the RBC refer to the location reference and orientation of the LRBG given in the same radio message.

The determination of the train position is always longitudinal along the route, even though the route might be set by the interlocking through a complex track layout, see Figs. 6.20 and 6.21 from Subset 026-3 [7].

The train position information defines the position of the train front in relation to a Balise Group, which is called Last Relevant Balise Group (LRBG).[7] Train position information, as defined in Subset 026-3 [7], see Fig. 6.22, includes:

- The estimated train front-end position by on-board, defined by the estimated distance measurement (from the odometry function) between the LRBG and the front end of the train;
- The train position confidence interval;
- Directional train position information in reference to the Balise Group orientation of the LRBG, regarding:
  - the position of the train front end (nominal or reverse side of the LRBG),
  - the train orientation (if there is an active cab,[8] this one defines the orientation of the train, i.e. the side of the active cab is considered as the front of the train. If

[7] Balise Groups, which are marked as unlinked, can never be used as LRBG.

[8] The active cab is the cab associated with an ERTMS/ETCS on-board equipment, from which the traction is controlled. The train orientation cannot be affected by the direction controller position.

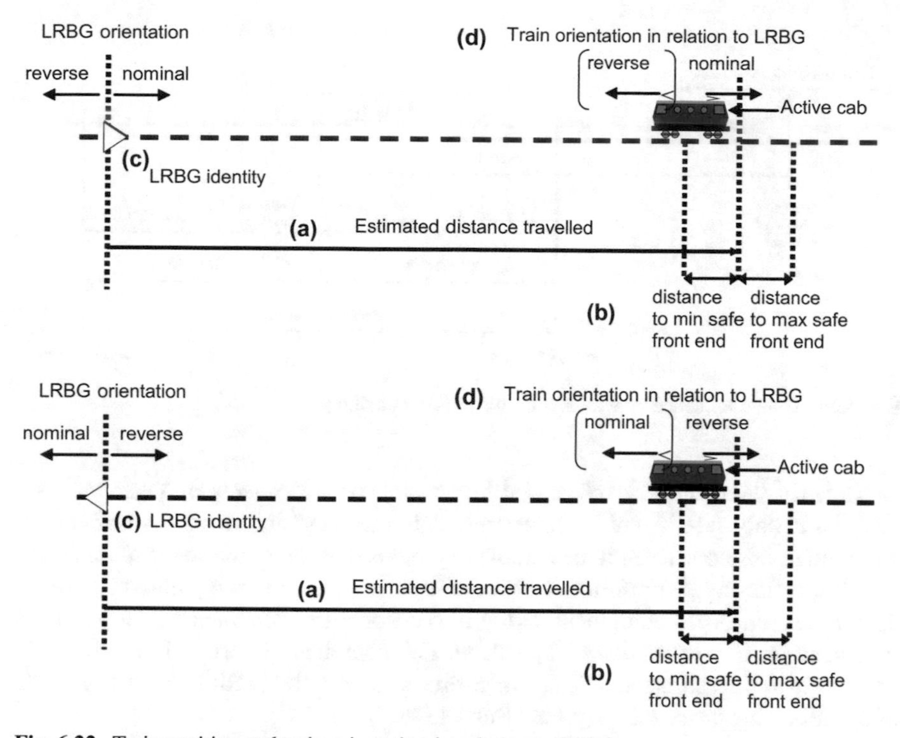

Fig. 6.22 Train position and train orientation in relation to LRBG

no cab is active, the train orientation is as when a cab was last active). The "train orientation relative to LRBG" is defined as the train orientation with respect to the orientation of the LRBG. It can be either "nominal" or "reverse",
  - the train running direction;
- A list of LRBGs, which may alternatively be used by trackside for referencing location dependent information.

In Fig. 6.23, as in Subset 026-3 [7], the train front-end position is identified by the on-board equipment in the following way:

- The estimated front-end position;
- The max(imum) safe front-end position, differing from the estimated front-end position by the under-reading amount in the distance measured from the LRBG plus the location accuracy of the LRBG (i.e. in relation to the orientation of the train this position is in advance of the estimated position);
- The min(imum) safe front-end position, differing from the estimated front-end position by the over-reading amount in the distance measured from the LRBG plus the location accuracy of the LRBG (i.e. in relation to the orientation of the train this position is in rear of the estimated position).

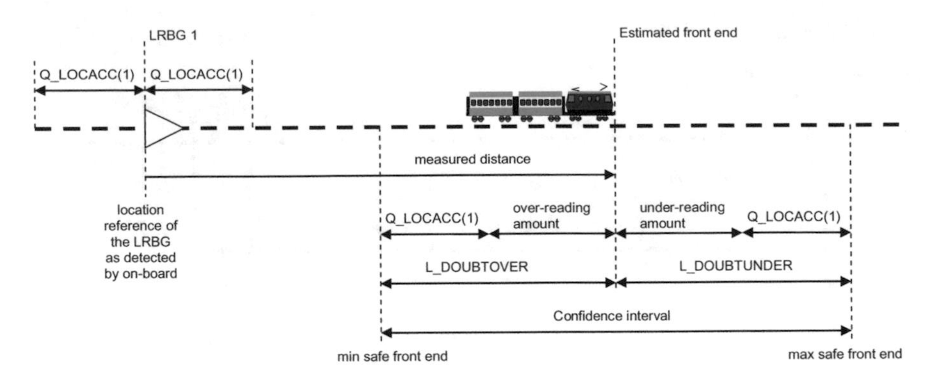

Fig. 6.23 Train confidence interval and train front-end position

The rear-end position is referenced in the same way. However, min safe rear-end position is only safe if sent together with train integrity[9] information (concept not analysed in this book to limit the complexity and is thus out of the scope of this text).

All location-related information transmitted from trackside equipment must be used by the on-board equipment taking into account the confidence interval of the train position, if required for safe operation. This confidence interval of the train position refers to the estimated distance measurement from the LRBG and is a function of the following terms (see Fig. 6.23 Subset 026-3 [7]):

- On-board over-reading amount and under-reading amount (odometer accuracy plus the error in the detection of the Balise Group location reference). The over-reading amount and the under-reading amount must be equal to or lower than (5m + 5% of the measured distance) = (4 m + 1 m + 5% of the measured distance) as stated in [20] where 1 m is the maximum location accuracy for vital purposes for each balise, when a balise has been passed [9];
- The location accuracy of the LRBG which corresponds to a fixed value estimated during the signalling design phase and verified by means of the measurement at the time of balise installation. This location accuracy is given by the Q_LOCACC value associated with the balise reference location of LRBG and received in the linking information, if available, or the corresponding national value, or the corresponding Default Value if the National Value is not applicable;
- Distance information received from trackside and related to the location data to be used on-board. Distance information is evaluated on-board as nominal information (without taking into account any tolerances, as tolerances must be included in the corresponding Q_LOCACC value).

Based on the odometry accuracy which is assumed and modelled as proportional to the train travelled distance, the train confidence interval increases in relation to the distance travelled from LRBG until it is reset when another linked Balise Group

[9]Train integrity is the level of belief in the train being complete and not having left coaches or wagons behind.

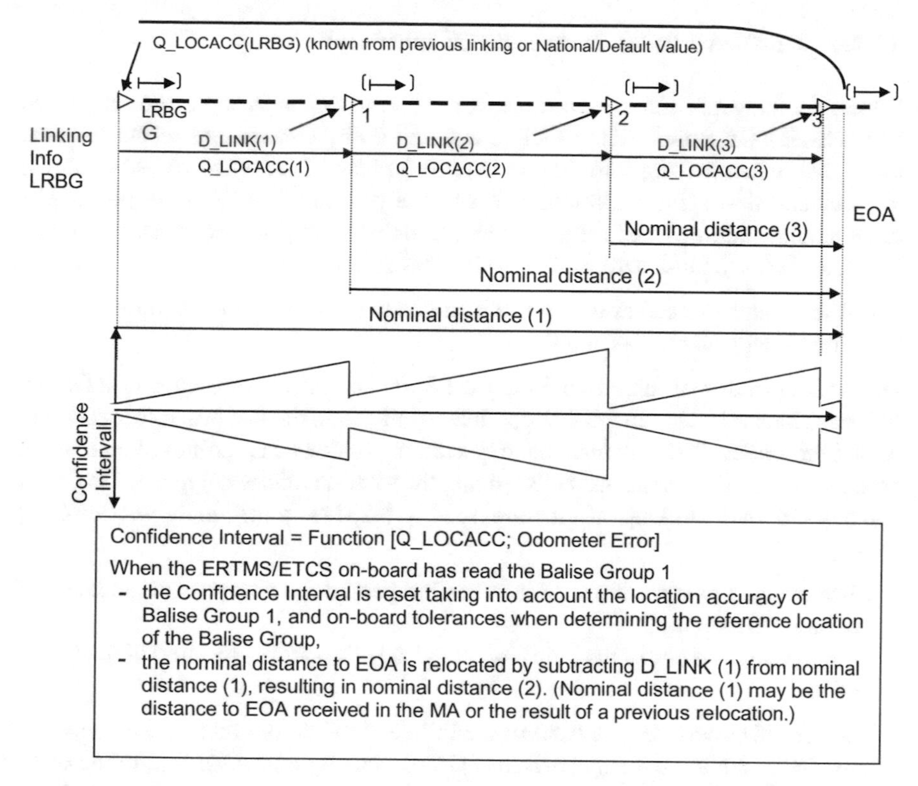

Fig. 6.24 Reset of the train confidence interval and relocation on change of LRBG

becomes the LRBG, see Fig. 6.24 as in Subset 026-3 [7]. When another Balise Group becomes the LRBG or when evaluating location-related trackside information, which is referred to a previously received Balise Group different from the LRBG, all the location-related information is relocated by subtracting one of the distance reported below from the distances that are counted from the reference Balise Group of the location-related information:

- The distance between the reference Balise Group of the location-related information and the LRBG, retrieved from linking information in case it is available and it includes both the reference Balise Group and the LRBG; or
- In all other cases, the estimated travelled distance between the reference Balise Group of the location-related information and the LRBG.

It is always the trackside responsibility to provide linking in due course or, where linking is not provided, the trackside can include provisions, if deemed necessary, during the signalling design phase for including such provisions in the distance information.

6.3.4.4 On-Board Rules for Accepting Linked Balises

When no linking information is used on-board (i.e. no Balise Group(s) are announced by trackside), On-Board ERTMS/ETCS takes all Balise Groups into account. On the other side, when linking information is used on-board (i.e. when Balise Group(s) are announced and the minimum safe antenna position[10] has not yet passed the expectation window of the furthest announced Balise Group, the expectation window is defined below), On-Board ERTMS/ETCS takes into account only:

- Balise Groups marked as linked and included in the linking information and
- Balise Groups marked as unlinked.

Due to the location accuracies of both the LRBG and the announced Balise Group, the space interval between the first possible location and the last possible location to accept the Balise Group defines the expectation window. The on-board equipment accepts a Balise Group (i.e. the balise giving the location reference) marked as linked and included in the linking information (see Fig. 6.23) as defined in Subset 026-3 [7] from

- when the max safe front end of the train has passed the first possible location of the Balise Group until
- the min safe front end of the train has passed the last possible location of the Balise Group

taking the offset between the front of the train and the balise antenna into account. The on-board equipment expects Balise Groups one by one according to the order given by linking information: it supervises only one expectation window at a time. The on-board equipment stops expecting a Balise Group and expects the next one announced in the linking information (if any) when:

- The Balise Group is found inside its expectation window;
- A linking consistency error is found due to reaching the end of the expectation window without having found the expected Balise Group or early reception of a Balise Group expected later.

If linking information is used, the on-board reacts according to the linking reaction information in the following cases, as in Subset 026-3 [7]:

- If the location reference of the expected Balise Group is found in rear of the expectation window;
- If the location reference of the expected Balise Group is not found inside the expectation window (i.e. the end of the expectation window has been reached without having found the expected Balise Group);
- If inside the expectation window of the expected Balise Group another announced Balise Group (i.e. another Balise Group included in the linking information), expected later, is found (to identify that another Balise Group has been found, the

[10]It is calculated by subtracting the distance between the active Eurobalise antenna of the BTM installed on the train and the front end of the train from the min safe front-end position.

on-board equipment has detected and processed at least one telegram sent by a balise of the other announced Balise Group).

The ERTMS/ETCS on-board equipment ignores (i.e. it does not consider as LRBG) a Balise Group found with its location reference outside its expectation window and, if the Balise Group is passed in the unexpected direction, rejects the message from the expected group and trips the train.

If the location reference balise of the linked Balise Group is duplicated and the on-board is only able to correctly evaluate the duplicating one, the duplicating one is used as location reference for the Balise Group, and it is found within the expectation window, no linking reaction is applied.

ERTMS/ETCS specification accurately describes the reactions to be applied by the on-board equipment when it detects a Balise Group message inconsistency for both linked or unlinked Balise Groups, see [7] for details. However, independently of the linking reaction set by trackside, if two consecutive linked Balise Groups announced by linking information are not detected and the end of the expectation window of the second Balise Group has been passed, the ERTMS/ETCS on-board commands the service brake and the driver is informed. At standstill, the location-based information (i.e. location data, this is the set of the data that refer only to specific locations) stored on-board shall be shortened to the current position determined by the train odometry.

6.3.4.5 Position Report to RBC

In ERTMS Level 2, the RBC, a computer-based platform, elaborates messages to be sent to the train on the basis of information received from external trackside systems and on the basis of information exchanged with the on-board subsystems. The main objective of these messages is to provide movement authorities to allow the safe movement of trains on the railway infrastructure area under the RBC responsibility. To this end, RBC acquires the position of each controlled train (i.e. the train position always refers to the front end of the respective engine with regard to the train orientation) by means of the train position report. This report contains at least the following position and direction data, see Fig. 6.22:

- The measured distance (by using the train odometry) between the LRBG and the train estimated front end;
- The distance from the estimated front-end position to the min safe (estimated) front-end position and the distance from the estimated front-end position to the max safe (estimated) front-end position;
- The identity of the location reference, the LRBG;
- The orientation of the train in relation to the LRBG orientation;
- The position of the front end of the train in relation to the LRBG (nominal or reverse side of the LRBG);
- The estimated speed;
- Train integrity information;
- Direction of train movement in relation to the LRBG orientation;

- Optionally, the previous LRBG.

The ERTMS/ETCS specification foresees one or a combination of the following different policies that an On-Board ERTMS/ETCS can adopt for sending the position report:

- Periodically in time;
- Periodically in space;
- When the max safe front end or min safe rear end of the train has passed a specified location;
- At every passage of a LRBG compliant Balise Group;
- Immediately.

Such a policy is commanded from the RBC and remains valid until a new command is received from the RBC. However, independently from the position report policy commanded by the RBC, the on-board equipment also sends a position report to the RBC if at least one of the following events occurs:

- The train reaches standstill, if applicable to the current ERTMS Operational mode (i.e. in short, mode);
- The mode changes;
- The driver confirms train integrity;
- A loss of train integrity is detected on-board;
- The train passes a RBC/RBC border with its min safe rear end;
- The train passes a RBC/RBC border with its max safe front end;
- The train passes a level transition border with its min safe rear end;
- The ERTMS level changes;
- A communication session is successfully established;
- The train passes a LRBG compliant Balise Group, if no (valid) position report parameters (i.e. no RBC command about the position report policy) are stored on-board.

6.3.4.6 Data for Safe Train Movement

To control the train movement in an ERTMS/ETCS-based system, the ERTMS/ETCS Trackside Subsystem provides information concerning both the route set for the train and the track description for that route to the ERTMS/ETCS on-board subsystem. This information is composed of [7]:

- Permission and distance to run, named the Movement Authority (MA), see below;
- When needed, limitations related to the Movement Authority, i.e. mode profile for On Sight, Limited Supervision or Shunting always sent together with the MA to which the information belongs;
- Linking information when available;
- Track description covering as a minimum the whole distance defined by the MA. Track description includes the following information:

- The Static Speed Profile (SSP)—It is a description of the fixed speed restrictions of a given piece of track, e.g. such speed restrictions can be related to the maximum line speed, the curves, the points, the tunnel profiles and the bridges.
- The gradient profile—It provides a gradient value for each location within the piece of track covered by the profile; the gradient value is identified as a positive value for an uphill slope, and with a negative value for a downhill slope.
- Optionally, Axle Load Speed Profile (ASP)—It is as a non-continuous profile where, for each section, the different speed value(s) for which minimum axle load category applies is specified. The ERTMS/ETCS on-board equipment considers the most restrictive speed restriction lower than or equal to that of the train.
- Optionally, speed restriction ensures a given permitted braking distance. This restriction is given by means of a non-continuous profile defining:
  The start and the end locations for the speed restriction;
  The permitted braking distance used to calculate the speed restriction value;
  Whether the permitted braking distance is to be achieved with the service brake or the emergency brake[11];
  A single gradient value applicable for the calculation.
- Optionally, track conditions used to inform the driver and/or the train of a condition in front of the train. A track condition can be given as profile data or location data. Examples of track conditions are:
  Powerless section, lower pantograph;
  Powerless section, switch off main power switch;
  Non-stopping area;
  Tunnel stopping area;
  Big metal masses, ignore on-board integrity check alarms of balise transmission;
  Radio hole, stop supervision of the loss of safe radio connection.
- Optionally, route suitability data that define which values concerning loading gauge, traction system and axle load category a train must meet to be allowed to enter the route. The route suitability data is sent as location data.
- Optionally, areas where reversing is permitted, i.e. areas, where initiation of reversing of movement direction is possible, i.e. change the direction of train movement without changing the train orientation.
- Optionally, changed adhesion factor to adjust the emergency brake model of the train.

[11]The emergency brake system provides a guaranteed maximum deceleration as stated by the train manufacture; the emergency brake command, activated by the on-board platform, can be released at standstill only. On the other end, the service brake system provides a non-safe brake system; the service brake command, activated by the on-board platform, can be released when the corresponding revocation condition is met. When the application of the service brake fails, the emergency brake command is normally issued.

It is the responsibility of trackside that (*a*) the on-board equipment has received the information valid for the distance covered by the Movement Authority and (*b*) the safe distance beyond the EOA/LOA is long enough to brake the train from the target speed to a standstill without any hazardous situation. On the other side, the on-board equipment is responsible for applying the brakes if no new information is received when the LOA is passed.

A Movement Authority includes the following information:

- The End Of Authority (EOA), which is the location to which the train is authorized to move;
- The target speed at the EOA, which is the permitted speed at the EOA; when the target speed is not zero, the EOA is called the Limit of Authority (LOA). This target speed at the EOA can be time limited;
- If no overlap[12] exists, the Danger Point, which is a location beyond the EOA that can be reached by the front end of the train without a risk for a hazardous situation. Examples of Danger Points are:
  - the entry point of an occupied block section (if the line is operated according to fixed block principles),
  - the position of the safe rear end of a train (if the line is operated according to moving block principles),
  - the fouling point of a switch, positioned for a route, conflicting with the current direction of movement of the train (both for fixed and moving block mode of operation);
- The end of an overlap (if used in the existing interlocking system), which is a location beyond the Danger Point that can be reached by the front end of the train without a risk for a hazardous situation. This additional distance is only valid for a defined time;
- A release speed, which is a speed limit under which the train is allowed to run in the vicinity of the EOA, when the target speed is zero. The release speed may be necessary in Level 2 for allowing a train to approach the EOA where the permitted speed reaches zero and might be too restrictive to permit acceptable driving due to inaccuracy of the measured distance. One release speed can be associated with the Danger Point and another one with the overlap. Release speed can also be calculated on-board the train;
- The timeouts associated with the possible sections whose the MA is composed of. The last MA section is called End Section. A timeout (named Section timeout) can

[12]The section of line in advance of a stop signal that must be unoccupied and, where necessary, locked before and during a signalled running movement to the rear of the signal to avoid an accident if the train brakes do not perform as expected and the train passes the End Of Authority. It is normally a piece of track beyond the Danger Point.

be associated with each section to be used for the revocation of the associated route when the train has not entered into it yet. In addition, a second timeout (named End Section timeout) can be attached to the End Section to be used for delaying the automatic route release when the train has entered the End Section; this delay makes sure that the train has come to a standstill before any switches inside the route can be moved.

In Level 2, the on-board equipment requests MA when the driver selects start and, based on the information received from RBC, can request a new Movement Authority in accordance with the following policy:

- A defined time before the train reaches the pre-indication location for the EOA/LOA assuming it is running at the warning speed;
- A defined time before the Section timer (not the End Section timer, not the Overlap timer) for any section of the MA expires, or before the LOA speed timer expires.

Together with the MA request, the on-board informs the RBC about the reason(s) for having requested MA (e.g. start selection by driver, achievement of the time instants associated respectively with the following events "time before reaching preindication location for the EOA/LOA" or "time before a section timer expires" or "the LOA speed timer expires", the track description has been deleted).

6.3.4.7 Emergency Messages

The RBC can send emergency messages individually to each on-board equipment. Each emergency message is acknowledged **at application level**, and this acknowledgement informs the RBC about the **use** of the emergency message by the on-board equipment. An emergency message can be individually revoked.

The RBC can stop a train with an unconditional or a conditional emergency stop message that contains the information of the new stop location referred to the LRBG. The on-board equipment immediately commands the train trip when it receives an unconditional emergency stop message. On the other hand, depending on when the conditional emergency stop message is received, the on-board equipment implements one of the following policies:

- When receiving this message, the train has already passed with its min safe front end the new stop location, the emergency stop message is rejected, and the RBC is informed.
- Otherwise, if the train has not yet passed with its min safe front end the new stop location, the emergency stop message is accepted; however, this location is

used by the on-board to define the new EOA/SvL only if not beyond the current EOA/LOA. The Supervised Location (SvL) is defined on-board as:

- the end of overlap[13] (if any and before timeout);
- if not, the Danger Point (if any);
- if not, the End Of Authority.

As long as a Limit of Authority is supervised, no SvL is defined on-board.

6.3.4.8 Speed and Distance Monitoring

Under the assumption that brake system, wheel/rail adhesion and brake characteristics are compliant with the declared performance values, the speed and distance monitoring is the supervision of the speed of the train versus its position to assure that the train remains within the given speed and distance limits from targets. In the context of the speed and distance monitoring, a target is defined by both a target location and a target speed to which the train must decelerate before reaching the target location. In order to meet this objective, the on-board equipment uses dynamically computed brake deceleration curves related to the supervised targets.

Railway brakes have a statistical behaviour, and braking distances vary within the typical distribution for a given condition. Therefore, correction factors are incorporated for the speed and distance monitoring. The on-board continuously supervises a list of targets, which may include:

- The locations corresponding to a speed decrease of the most restrictive speed profile (MRSP)[14] (if any), which are in advance of the max safe front end of the train. This target is removed from the list when the max safe front end of the train has passed such a target location;
- The Limit Of Authority (LOA);
- The End Of Authority (EOA) and the SvL, if the target speed at the EOA is equal to zero;
- The location deduced from the maximum permitted distance to run in Staff Responsible, with a target speed zero.

The list of supervised targets is re-evaluated when any of the elements of the list (i.e. a supervised target) is changed. With respect to the supervision speed limits, the On-Board ERTMS/ETCS monitors the following limits:

- Emergency brake intervention (EBI);

[13]It is common practice in many railway and metro systems to use overlap in the entry route. Overlap provides additional free space beyond the stopping point, with guarantee that no movement is allowed in that area. The purpose of overlap is to allow easy stop of train at platform and improve safety of operation; a typical case is in stations where platform is almost as long as trains, arriving train may exceed stopping point and overlap guarantees that this shall cause no safety issue.

[14]The MRSP is a description of the most restrictive speed restrictions the train must obey on a given piece of track.

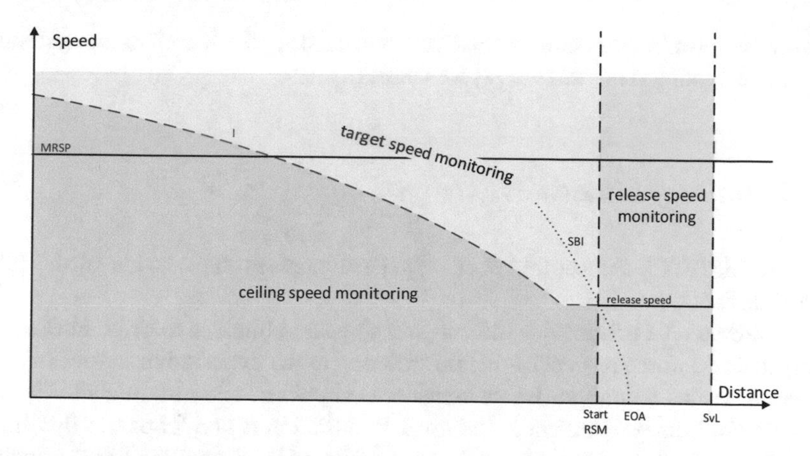

Fig. 6.25 Different types of speed and distance monitoring

- Service brake intervention (SBI);
- Warning (W);
- Permitted speed (P);
- Indication (I);
- Pre-indication location;
- Release speed monitoring start location.

The purpose of the emergency brake intervention supervision limit is to assure that the train remains within the various limits (in distance/speed) imposed by the trackside. The purpose of all other supervision limits is to assist the driver in preventing an emergency brake intervention by maintaining the speed of the train within the appropriate limits. By comparing the train speed and position to the supervision limits defined above, the on-board equipment generates braking commands, traction cut-off commands and relevant information to the driver. The following types of speed and distance monitoring are defined in ERTMS/ETCS, see Fig. 6.25:

- Ceiling speed monitoring (CSM);
- Target speed monitoring (TSM);
- Release speed monitoring (RSM).

Ceiling speed monitoring is the speed supervision in the area where the train can run with the speed as defined by the MRSP without the need to brake to a target. Target speed monitoring is the speed and distance supervision in the area where the specific information related to a target is displayed to the driver and within which the train brakes to a target. Finally, release speed monitoring is the speed and distance supervision in the area close to the EOA where the train is allowed to run with allowed release speed (i.e. a speed not equal to zero) to approach the EOA.

Once a train interface command is triggered (i.e. traction cut-off, service brake or emergency brake) in accordance with the triggering criteria foreseen in ERTMS/ETCS specification (e.g. based on the minimum safe front end or the maximum safe front end or the minimum safe rear end), the on-board applies it until its corresponding revocation condition is met (see [7] for details).

6.3.5 Railway Mission Profile

The ERTMS/ETCS Subset-088 Part 0 [21] summarizes the mission of the ETCS system as follows:

"To provide the driver with information to enable him/her to drive his/her train safely and to enforce respect of this information to the extent advised to ETCS."

Based on this definition, let us consider a simple mission of an ERTMS Level 2 fitted train from going from a Station A to a Station B and let us use this simple mission as an example for showing the application of the Signalling Basic Principles described in Sect. 6.3.4.

For simplicity, the example assumes that both stations and track lines are already in a Level 2 track area and the on-board equipment is not equipped with the Cold Movement Detection (CMD) function; CMD allows for (*a*) the detection and recording of vehicle movements, while the ERTMS on-board equipment is in No Power (i.e. switch off) for a duration of at most 72 h (note that the ERTMS on-board equipment cannot determine train movement when unpowered) and (*b*) then, if there are the conditions, the automatic revalidation of the train position information when the ERTMS on-board equipment is again powered on. The on-board platform configuration without the CMD function is the worst-case configuration with respect to train localization during the Start of Mission because the on-board ETCS cannot validate its position after the power on.

When the driver powers on the train, the ERTMS/ETCS equipment automatically initiates the default Stand-By (SB) ERTMS Operation Mode. If no safety fault is detected during and at the completion of the auto tests, the SB Operation Mode is maintained, otherwise the transition to the System Failure (SF) Operation Mode occurs. In SF, the ERTMS/ETCS on-board equipment permanently commands the emergency brake. On the other hand, in Stand-By mode, it performs the Standstill Supervision; i.e., it automatically stops any train movement that exceeds a distance specified by the national/default value. In SB, the desk of the engine can be opened (i.e. it becomes active) or closed. No interaction with the driver is normally possible as long as the desk is closed.

In order to allow the train starting its movement, the driver has to open the desk and to perform the procedure Start of Mission (SoM) [Subset026-5]. The steps of this procedure depend on the availability on-board of the following data:

- Driver ID;
- ERTMS/ETCS level;
- RBC ID/phone number;
- Train data;
- Train running number;

- Train position.

and of their status:

- "Valid" (the stored value is known to be correct);
- "Invalid" (the stored value may be wrong);
- "Unknown" (no stored value available).

In the context of the Start of Mission of this example, let us suppose that the train is located near the START marker of the station platform (e.g. this START marker is 30 m before the main signal of the platform in Italy, and a BG composed of two balises is associated with such a marker), and the driver has started the SoM procedure by entering the driver ID and the train running number, and by revalidating the ERTMS Level (e.g. Level 2). Once the driver revalidates Level 2, the on-board platform contacts the RBC identified by means of the further data entered by the driver or of the related data already available on-board.

For the sake of simplicity, let us assume that the communication session with RBC is successfully open. Then, the on-board platform sends the RBC the SoM Position Report with train position set to invalid or unknown owing to the unavailability of the CMD function. In addition, let us suppose that (*a*) the RBC cannot validate the train position because it does not have safe information from other trackside equipment and (*b*) the RBC also accepts the train. Then, the RBC informs the ERTMS/ETCS on-board equipment that the train is accepted without valid position data.

After this handshake between the RBC and the on-board equipment, the on-board equipment allows the driver to continue the SoM by performing the train data entry. Once the data entry phase is completed, the on-board equipment sends the train data to the RBC for the RBC checking and validation. Let us assume that the RBC successfully validates the train data, thus enabling the on-board platform to offer the driver the possibility to select "Start". After having received the authorization to complete the SoM procedure, the driver presses start, and this initiates a request for a Movement Authority to the RBC. In Level 2 with invalid/unknown position, the RBC cannot provide a MA. A possible way to allow train movements is the RBC to authorize a mode change that allows the driver to move the train under his own responsibility in ERTMS/ETCS areas; the authorized mode must allow the on-board equipment to supervise the train movement against a nationally defined ceiling speed and, if available, a maximum distance to run. However, this change of mode must not be assumed to be authority to move; this authorization to move must be given as permission to proceed received from the signaller or the manager responsible for controlling the train movement, or by the clearing of a signal or lineside indicator showing a proceed aspect.

Let us suppose that the RBC has authorized the SR mode with the default ceiling speed equal to 30 km/h and maximum distance set to infinite. Once the train has started to move and has passed over the START Balise Group, the on-board equipment sends the train position report to RBC (see Sect. 6.3.4.5). RBC performs its own check and, for example, sends the Track Ahead Free (TAF) request to the driver that acknowledges it. This acknowledgement enables the RBC to send the FS MA.

Note that a BG (e.g. named "IMPERATIVO" in Italy) is associated with the main signal and this BG contains the command "Stop if in SR Mode". Therefore, if for any reason, the on-board reads the "IMPERATIVE" BG before having implemented the mode transition from SR to FS, the on-board will immediately command the emergence brake. This engineering rule guarantees a SIL 4 SoM procedure at the station.

For the sake of completeness, let us also provide the example of SoM executed in line. To start the SoM procedure, the driver needs a special authorization from the signalling manager. Suppose that this SoM is done with "Invalid or Unknown" on-board stored information. As done for the SoM procedure in station, the driver has to enter/revalidate the driver ID, train running number, the ERTMS level and the data for allowing the connection with the RBC. Let us assume that the communication session is successfully open, the on-board has successfully sent the SoM Position Report with "Invalid or Unknown" position data to the RBC, the RBC has accepted the train, has validated the train data, and has enabled the driver to push START, and, after having received the authorization, the driver has pressed the START button. As the RBC cannot assign a FS MA, it authorizes the on-board mode change from SB to SR. Once the train has been authorized and started to move, it continues its run at a maximum speed equal to 30 km/h. For each radio block section of the line, there is at least one BG. When it passes the first BG (for simplicity, composed of two balises) that it encounters, it sends the BG position report to the RBC.

In principle, by using this position report, the RBC might assign the FS MA to this train. However, as a MA must only be issued once the ERTMS system has proved that all of the conditions have been met for guaranteeing a safe train movement (i.e. the route has been set and locked, and the conditions for a non-permissive or permissive route have been met), national signalling procedures require the successful execution of additional defence checks on the train position. In the context of this example, the driver continues its run in SR mode up to the first stop Track Ahead Free (TAF) lineside marker (i.e. the location where the RBC requests to driver to confirm that the track between the head of the train and the next signal or board marking signal position is free). Therefore, the driver stops the train at this location and the on-board equipment sends the position report at standstill. When the RBC receives this position report, it checks that (*a*) the train speed is zero (i.e. the train can be considered in the standstill condition) and (*b*) the on-board measured distance from the previous detected Balise Group along with the confidence associated with such a measurement is coherent with the distance measured by RBC based on the RBC track database. Note that the TAF window must have a length suitable for coping with these uncertainties and the planned margin foreseen at the level each Generic/Specific ERTMS application; for example, this TAF window has a length equal to 100 m in Italy. If these checks are successfully passed, the RBC sends the TAF request to the on-board equipment. If the driver confirms (by responding to an on-board TAF request) that there is no train/vehicle between the train and the signal or ERTMS stop marker ahead, then the RBC sends the MA to on-board equipment, e.g. the Full Supervision (FS) MA. From now on, the on-board equipment will use the data contained within the received MA to calculate the braking curves and the permitted, warning and intervention speed curves. The train movements are supervised with respect to such braking curves. Information about the ERTMS MA will be displayed on the ERTMS DMI, and the driver must control the train in accordance with the information displayed.

These examples have been used for providing a high-level overview of the Start of Mission procedure, based on some national signalling procedures, under non-severe conditions and for very simple missions. However, note that, in order to carry out a complete and accurate safety analysis of ERTMS systems, the Railways Standard CENELEC 50126 [22] requires the definition of a Mission Profile to outline the expected range and variation in the mission with respect to parameters such as time, loading, speed, distance, stops and tunnels in the operational phases of the life cycle. To this end, the verification and validation phases of **interoperable** ERTMS systems have used the Mission Profile described in Subset 088 [23] that has played a critical role in the design, development and safety analysis of the ERTMS/ETCS because targets for the design of equipment are derived from it.

The reference Mission Profile [23] is related to both high-speed and conventional line applications and has been agreed with representatives of the European Railway Authorities to determine realistic exposure times that a passenger on a train might experience in the defined one-hour journey.

This Mission Profile consists of two parts as follows:

- The reference infrastructure, see Table 6.3;
- The operational parameters, see Table 6.4.

As the Mission Profile has also been used for carrying out the safety analysis, the parameters with asterisk (*) denote those parameters that have been used in the safety apportionment process (see Sect. 6.3.6).

6.3.6 Main Dependability Requirements

This paragraph summarizes the main ERTMS/ETCS performance and safety requirements. As the balise is the interoperable constituent used in ERTMS/ETCS for location referencing, let us start with the relevant requirements applicable to the Eurobalise Transmission System [1, 9].

The Eurobalise Transmission System is a safe spot transmission-based system conveying safety-related information from the trackside infrastructure to the on-board equipment. This system consists of the (trackside) Balise and the On-board Transmission Equipment, i.e. the Balise Transmission System (BTM) and the related antenna (that is part of the ERTMS/ETCS on-board constituent), see Fig. 6.26.

In ERTMS/ETCS, balises are of either fixed type or controlled (switchable) type. However, as the concept of Virtual Balise based on the GNSS technology (see next Chap. 7) is applicable to the fixed type balises only, requirements and the related analysis associated with controlled balises will not be addressed in the rest of this book.

Table 6.3 Mission Profile—reference infrastructure

Parameter description	Value	
	High-speed rail	Conventional rail
Length of the line travelled in one hour	260 km/h	80 km/h
Number of Radio Block Centres	$3\,h^{-1}$	$1\,h^{-1}$
Number of station (general) and/or stopping points	$25\,h^{-1}$	$25\,h^{-1}$
Number of stations (stations where Start of Mission is implied due to awakening of the train)	$1\,h^{-1}$ (*)	$2\,h^{-1}$ (*)
Number of changes in direction of travel (where Start of Mission is implied)	$1\,h^{-1}$ (*)	$2\,h^{-1}$ (*)
Number of tunnels	$10\,h^{-1}$	$3\,h^{-1}$
Number of trains on the line	$15\,h^{-1}$	$15\,h^{-1}$
Number of signals (0 possible for level 2)	$0–200\,h^{-1}$	$0–50\,h^{-1}$
Maximum distances between Balise Groups	2.5 km	2.5 km
% of journey with the maximum distance between Balise Groups	10%	10%
Number of unlinked Balise Groups (marked as Unlinked)	1 in 1000 (*)	4 in 1000 (*)
Number of repositioning Balise Groups (only Level 1)	1 in 100	1 in 100
Number of level transitions (including NTC X - NTC Y transitions)	$2\,h^{-1}$ (*)	$2\,h^{-1}$ (*)
Number of temporary shunting areas with number of border balises	1/66	1/66
Number of fixed shunting areas (after which Start of Mission is implied)	$1\,h^{-1}$ (*)	$1\,h^{-1}$ (*)
Number of national border transition	$1\,h^{-1}$	$1\,h^{-1}$

Table 6.4 Mission profile—operational parameters

Parameter description	Value	
	High-speed rail	Conventional rail
Average speed of trains of the line	260 km/h	80 km/h
Max. speed of trains of the line	350 km/h	250 km/h
Frequency of Balise Group messages	150–650 h^{-1} (*)	50–150 h^{-1} (*)
Frequency of Balise Group messages used only for reset of confidence interval (%), thus having a link reaction marked as no reaction	~90% (L2) (*) ~50% (L1) (*)	~90% (L2) (*) ~50% (L1) (*)
Frequency of radio messages track to train	100–360 h^{-1}	25–360 h^{-1}
Frequency of radio messages train to track	100–650 h^{-1}	50–650 h^{-1}
Frequency of emergency messages (only level 2)	4×10^{-4} h^{-1}	4×10^{-4} h^{-1}
Number of train data entry procedure, see Note A	2 h^{-1} (*)	4 h^{-1} (*)
Number of RBC/RBC Transitions	3 h^{-1}	1 h^{-1}
Max. expected loss of train integrity	N/A	N/A
Mean down time of a failed ETCS on-board balise receiver in an unfitted area	1 h (*)	1 h (*)
Mean down time of a non-detectable Balise Group	24 h (*)	24 h (*)

In the context of the Eurobalise Transmission System, the information is transmitted to the train only when the Antenna Unit of the On-board Transmission Equipment passes or stands over the corresponding balise; therefore, it is a spot transmission that occurs at the discrete location where the balise is installed. When a balise is detected, the Eurobalise Transmission System must (*a*) evaluate the location of the detected balise, using the available time and odometer information received from the ERTMS/ETCS Kernel [1], and (*b*) provide the balise information and the estimated balise location information to the ERTMS/ETCS Kernel. The estimated balise location must include the error in the time and odometer information.

Table 6.5 summarizes the main ERTMS/ETCS dependability requirements.

Subset 088 [23] describes the safety analysis carried out on an ERTMS/ETCS system based on the ETCS SRS [7] and the ETCS Reference Architecture reported in Fig. 6.12 when used in Application Levels 1 and 2. This analysis was performed

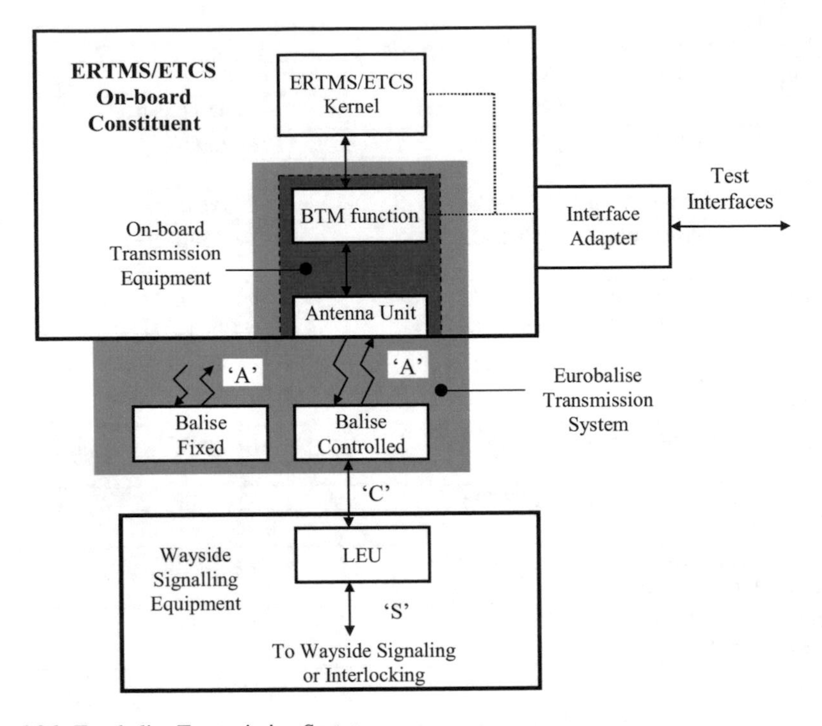

Fig. 6.26 Eurobalise Transmission System

taking into account both the ETCS functions and those functions external to the reference architecture that must be harmonized to enable interoperability. The output of this analysis was the safety requirements for ETCS Application Levels 1 and 2 that must be satisfied to guarantee the required technical interoperability.

In accordance with provisions of the CCS TSI [14] and with the Mission Profile defined in Tables 6.3 and 6.4, the maximum allowed rate of occurrence of the ETCS Core Hazard is 1.0×10^{-9}/h for ETCS on-board and 1.0×10^{-9}/h for ETCS trackside.

Based on (*a*) the ETCS architecture that foresees an on-board equipment, a trackside equipment and the airgap between them for transferring information from trackside to on-board and vice versa, and (*b*) the equal values of THR for on-board and trackside ETCS equipment, the top-level ETCS THRs have been apportioned as indicated in Fig. 6.27 [23], where the terms $THR_{On\text{-}board}$, $THR_{Trackside}$ and $THR_{Transmission}$ represent the numerical safety requirement respectively for the purely on-board, trackside and transmission functions.

As the transmission functions are actually carried out by either the on-board or trackside equipment, $THR_{Transmission}$ has also been partitioned into two equal contributions, one for the on-board equipment and the other for the trackside equipment (see Fig. 6.27). In order to perform a safety analysis and a more detailed THR appor-

Table 6.5 Main ERTMS/ETCS dependability requirements

Requirement Description	Note
Balise location reference accuracy—the balise location reference accuracy for vital purposes (meaning with a confidence level as that of SIL 4) must be within +/−1 m for each Balise, **when a balise (i.e. the location reference) has been passed**	Many ERTMS engineering and dimensioning rules [19] have been set up taken this balise location accuracy into account. For example, Sect. 4.1.1.4 of [19] states the minimum distance between the balise group and the EOA/LOA, when the last encountered balise of the Balise Group gives the order to Stop if in SR (this is a safety command for avoiding the access to an area when the responsibility of the train movement is only assigned to the driver); this minimum distance is 1.3 m plus the distance the train may run during the time required for transferring the uplink data to the ERTMS/ETCS Kernel (for train speeds lower than 80 km/h, this transferring time is equals to 100 ms). Therefore, the introduction of the Virtual Balise Concept, based on GNSS as described in the next chapter, will imply the review of the ERTMS engineering and dimensioning rules and of the mission profile because the high demanding location accuracy associated with the detection of Physical Balises is not valid any more for Virtual Balises
Balise Group orientation— The Eurobalise Transmission System must provide information to the ERTMS/ETCS Kernel that enables the ERTMS/ETCS Kernel to evaluate the direction of the train on the basis of the reported sequence of passed Balises	
Cross-talk protection— The Eurobalise Transmission System must not allow a valid telegram to be passed through, from a Balise located in a cross-talk protected zone, to the On-board ERTMS/ETCS Kernel. The cross-talk protected zone is the zone in the vicinity of the Balise where transmission is not intended to take place	See Appendix A.2 (Cross-Talk Protected Zone)

Table 6.5 (continued)

Requirement Description	Note
Response time The delay between receiving of a balise message and applying the command requested in the balise message or the reaction associated with Balise Group must be less than 1 s. The delay between receiving of a balise message and initiating a communication session establishment, as requested by the message contained in the Balise Group, must be less than 1.5 s	For details, see Sect. 5.2.1.1. of [20]
Accuracy of distances measured on-board—For every measured distance s the accuracy shall be better or equal to ±(5m + 5% s), i.e. the over-reading amount and the under-reading amount must be equal to or lower than (5 m + 5% s), where this performance requirement includes the error for the detection of a balise location once passed	For details, see [19]
Management of faults and failures—If the On-board Balise Transmission Equipment is not able to detect balises, this condition must be reported to the ERTMS/ETCS Kernel	For details, see Sect. 4.1.5 of [9]
Schedule adherence—The probability of having delay **caused by ERTMS/ETCS failures** must not be greater than 0.0027. The allowed average delay per train **due to ERTMS/ETCS failures**, at the end of an average trip of duration of 90 min, must not be greater than 10 min	Schedule adherence is the ability of a railway system of complying with the schedule of train running. A train is considered delayed when its delay exceeds 1 min. For details, see Sect. 2.2.2.1.1 of [24]

Table 6.5 (continued)

Requirement Description	Note
Operational availability—The operational availability of the ERTMS/ETCS, **due to all the causes of failure**, must not be less than $A_0 = 0.99973$ The ERTMS/ETCS quantifiable contribution to operational availability, **due to hardware failures and transmission errors**, must not be less than $A_0 = 0.99984$	For details, see Sect. 2.2.2.1.2 of [24]. Downtime, expressed in hours per year, is equal to $(1 - A_0) * 8760$
Immobilising failure—No one single fault must cause immobilising failures. When redundancies are utilized to prevent single failures to cause immobilising failures, appropriate measures which guarantee the independence of the redunded equipment must be adopted and documented. The Mean Time Between Immobilising hardware Failures MTBFIONB, **defined for on-board equipment**, must not be less than $2.7 * 10^6$ h. The Mean Time Between Immobilising hardware Failures MTBFITRK, **defined for Trackside Centralized equipment**, must not be less than $3.5 * 10^8$ h	An Immobilising Failure is an ERTMS Control/Command failure which causes the system to be unable to safely control two or more trains [24]: this is, a failure that causes two or more trains to be switched in on sight mode. The relevant mission is then defined as the ERTMS/ETCS operation in absence of Immobilising Failures

Table 6.5 (continued)

Requirement Description	Note
Service failure—The Mean Time Between Service hardware Failures MTBFSONB, **defined for on-board equipment**, must not be less than $3.0 * 10^5$ h. The Mean Time Between Service hardware Failures MTBFSTRK, **defined for trackside centralized equipment**, must not be less than $4.0 * 10^7$ h	A Service Failure is an ERTMS Control/Command failure which causes the nominal performance of one or more trains to be reduced and/or the system to be unable to safely control at most one train In the ERTMS/ETCS context, Service Failures can be identified as all the ERTMS/ETCS failures which cause the nominal performance of one or more trains to be reduced and/or at most one train to be switched in on sight mode. The relevant mission is then defined as the ERTMS/ETCS operation in absence of Service Failures [24]
Minor failure—The Mean Time Between Minor hardware Failures $MTBF - M_{ONB}$, **defined for on-board equipment**, must not be less than $8.0 * 10^3$ h. The Mean Time Between Minor hardware Failures $MTBF - M_{TRK}$, **defined for trackside centralized equipment**, must not be less than $1.0 * 10^5$ h	A Minor Failure is a failure which results in excessive unscheduled maintenance and cannot be classified as Immobilising or Service Failure. The relevant mission is then defined as the ERTMS/ETCS operation in absence of Minor Failures [24]

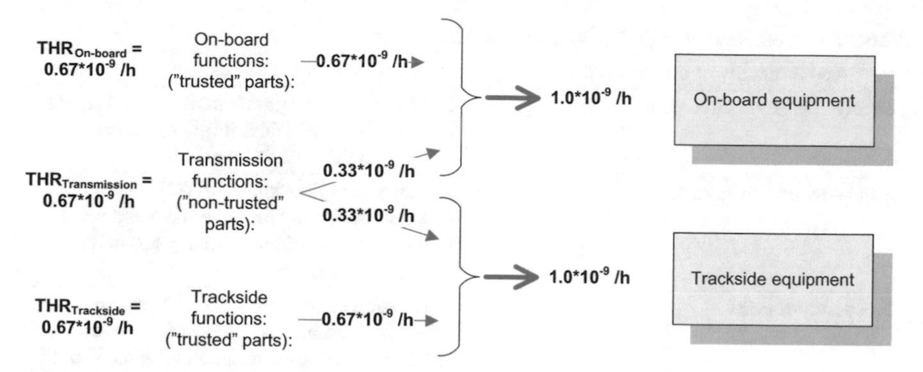

Fig. 6.27 High-level apportionment of top-level ETCS THRs to ETCS equipment

tionment that does not depend on a specific ETCS interoperable constituent and its implementation, the following approach has been adopted:

- For each Level 1 and Level 2, a system wide, generic, non-mandatory functional fault tree based on the failure modes of ETCS functions (see [7]) has been defined. The fault tree base events represent the low-level functions and data items of ETCS. This functional fault tree puts the function failure modes into a hierarchy leading up to the ETCS Core Hazard "Exceedance of the safe speed or distance as advised to ETCS". The hierarchy is generic and not mandatory for a product design;
- By using a bottom-up approach and starting from the functional fault tree, a detailed Fault Mode and Effects Analysis (FMEA) has been carried out (*a*) to establish all the possible mitigated events and the mitigation techniques and (*b*) to assign a criticality to each fault tree base event, see Table 6.6;
- In order to perform the quantitative apportionment of the ETCS THR to ETCS constituents, some fault tree base events have been further decomposed to a lower level to enable its assignment to the on-board or the trackside or the air gap between the on-board and the trackside. The apportionment process to define THR targets to be allocated to equipment and to specific functions for guaranteeing technical interoperability has also taken into account (*a*) the operational aspects, (*b*) the protective features **inherent in the design of ETCS** [7, 23] and (*c*) the frequency of occurrence of operational events in accordance with the reference Mission Profile.

Figure 6.28 represents the graphical representation of the identified ETCS hazardous events whose complete descriptions are reported in [25]. The occurrence of any one of the defined hazards must not lead to ETCS exceeding the top-level THR targets.

The parts coloured in red in Fig. 6.28 respectively outline the Eurobalise and the Euroradio Transmission Systems, whose transmission functions are active in ERTMS Level 2. The ETCS hazardous events associated with such transmission systems are:

- TRANS-BALISE 1 (Corruption);

Table 6.6 Definition of criticality of fault tree base events

Assigned criticality of base events	Interpretation of the assignment
Safety critical function/data	Safety critical function/data A function or data item of ETCS which, if it failed, would lead directly to the ETCS Core Hazard
Safety-related function/data	A function or data item of ETCS which if failed in addition with other independent functions or conditions could result in the ETCS Core Hazard
Not safety-related	A function or data item of ETCS which if failed in addition with other independent safety-related functions or conditions would not result in the ETCS Core Hazard

- TRANS-BALISE 2 (Deletion);
- TRANS-BALISE 3 (Insertion);
- TRANS-OB/RADIO-1 (Corruption);
- TRANS-OB/RADIO-2 (Deletion);
- TRANS-OB/RADIO-3 (Insertion);
- TRANS-TS/RADIO-1 (Corruption);
- TRANS-TS/RADIO-2 (Deletion);
- TRANS-TS/RADIO-3 (Insertion);

The TRANS-BALISE-n, TRANS-OB/RADIO-n and TRANS-TS/RADIO-n events refer only to errors occurring in the communication channel including the non-trusted [26] parts of transmitting and receiving entities.

The FMEA associated with these TRANS events [25] is reported in Table 6.7 [23, 25].

The safety hazard analysis carried out on on-board functions has also taken into account the possible effects associated with elements external to the on-board platform such as:

- Wrong engineering data;
- Incorrect installation of equipment;
- Incorrect data entered by the driver;
- Incorrect information at the train interface.

Such safety analysis has confirmed $THR_{\text{On-board}} = 0.67 \times 10^{-9}$ dangerous failures/hour for the pure on-board functions (i.e. inclusive of the parts of the hazard that arise due to failures inside the trusted parts of the on-board Eurobalise Transmission System and the on-board Euroradio Transmission System, and not inclusive of the parts associated with the not-trusted transmission functions performed in the on-board equipment).

On the other hand, the safety hazard analysis performed on trackside functions has also taken into account the possible effects associated with elements external to the trackside platform such as:

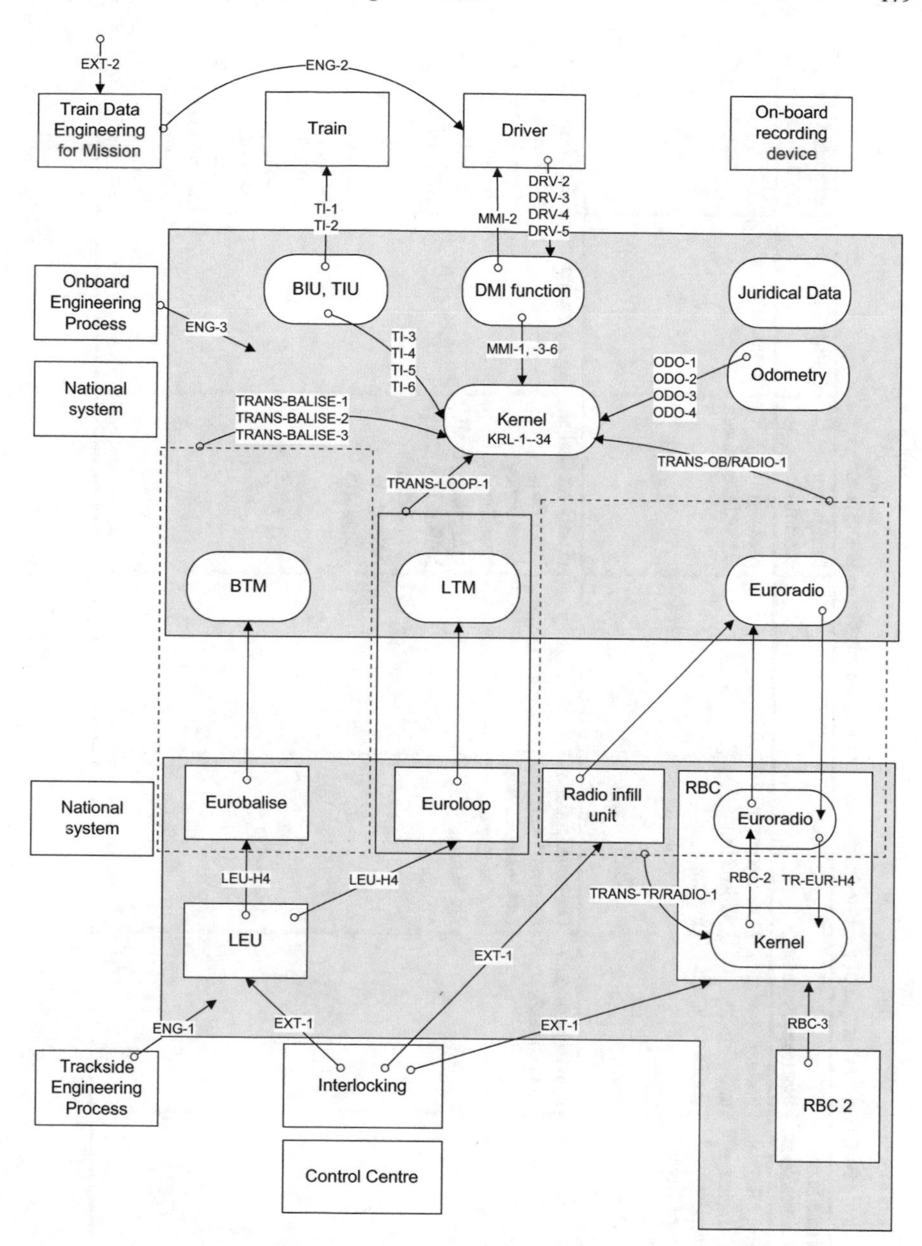

Fig. 6.28 Graphical representation of ETCS hazardous events

Table 6.7 FMEA related to the fault tree base events of the Eurobalise and the Euroradio transmission systems

Fault tree base event	Fault tree base event description	Affected ETCS functions or data	Explanation	Mode	Mitigation condition	Criticality	Exported conditions
TRANS-BALISE-1	Incorrect Balise Group message received by on-board Kernel functions as consistent. (corruption)	Provision of data to on-board (balise message)	Corruption of Balise Group message	All	Message consistency check	Safety critical	
TRANS-BALISE-2	Balise Group not detected by on-board Kernel functions (deletion)	Provision of data to on-board	On-board fails to receive data from balise and failure to detect any of the balises in the group	All	If only one balise is missed, consistency checking is mitigation. If all balises in a group are missed, linking is mitigation	Safety critical	The criticality of this failure is dependent upon the information missed within the unlinked Balise Group. Having two (or more) balises in the group can mitigate the hazard of deletion. In situations where deletion is critical, single Balise Groups are not appropriate

Table 6.7 (continued)

Fault tree base event	Fault tree base event description	Affected ETCS functions or data	Explanation	Mode	Mitigation condition	Criticality	Exported conditions
TRANS-BALISE-3	Inserted Balise Group message received by on-board Kernel functions as consistent. (insertion)	Provision of data to on-board (balise message)	Cross-talk of Balise Group message	All	Message consistency check. Balise Group linking	Safety critical	
TRANS-OB/RADIO-1	Incorrect radio message received by the on-board Kernel functions as consistent. (corruption)	Provision of data to on-board (MA data etc.)	Incorrect data includes corruption, late, repeated, etc.	All	Message consistency check has to fail. Messages are key coded to ensure authenticity and contain a timestamp to check sequencing and delay	Safety critical	Emergency messages are not covered by the MAC code and therefore there is no mitigation. If the on-board can decode the message as an emergency message the message will be acknowledged by the on-board to the RBC

Table 6.7 (continued)

Fault tree base event	Fault tree base event description	Affected ETCS functions or data	Explanation	Mode	Mitigation condition	Criticality	Exported conditions
TRANS-OB/RADIO-2	Radio message not received by the on-board Kernel functions (deletion)	Provision of data to on-board (MA data etc.)	Deletion in the communications channel resulting in the on-board being unable to receive a more restrictive MA. The on-board will be unable to receive emergency messages, no protection afforded against the loss of conditional emergency messages. However, this is conditional upon an emergency message being transmitted to the train	All	Train should be shortened via co-operative MA shortening. For shortening by an emergency message mitigation is provided by the radio link supervision. This ensures messages are received no later than a specified time (T_NVCONTACT) which should be limited to a safe default value defined by each railway. Section timeouts will also provide mitigation	Safety critical	
TRANS-OB/RADIO-3	Inserted radio message received by the on-board Kernel functions as consistent. (insertion)		Erroneous MA received by the on-board resulting in an exceedance of speed/distance	All	Message sequencing, time stamping and addressing as recommended by CENELEC 50159 render this event as non-hazardous	Safety critical	

Table 6.7 (continued)

Fault tree base event	Fault tree base event description	Affected ETCS functions or data	Explanation	Mode	Mitigation condition	Criticality	Exported conditions
TRANS-TS/RADIO-1	Incorrect radio message received by RBC Kernel functions as consistent. (corruption)	Provision and revocation of emergency messages. Provision of data to the on-board		All	Message consistency check Messages are key coded to ensure authenticity and time stamped as per CENELEC 50129-2 to check sequence and delay	Safety critical	
TRANS-TS/RADIO-2	Radio message not received by the RBC Kernel functions (deletion)		Loss of train reports and/or message acknowl-edgements	All	Train retains existing MA. Protection is afforded at the application level with transmission repeats	Safety critical	
TRANS-TS/RADIO-3	Inserted radio message received by the RBC Kernel functions as consistent. (insertion)				Message sequencing, time stamping and addressing as recommended by CENELEC 50159 render this event as non-hazardous	Safety critical	

- Wrong engineering data used in the preparation of a signalling scheme;
- Incorrect installation of equipment;
- System maintenance.

This analysis has confirmed $THR_{Trackside} = 0.67 \times 10^{-9}$ dangerous failures/hour for the pure trackside functions (i.e. inclusive of the parts of the hazard that arise due to failures inside the trusted parts of trackside Euroradio Transmission System, and not inclusive of the parts associated with the not-trusted transmission functions performed in the trackside equipment). This THR does not include any contribution associated with the cooperation between the trackside equipment and external interlockings because the effects of this cooperation are outside the ETCS responsibility.

As far as the safety analysis related to the transmission functions is concerned, detailed fault tree events associated with transmission and reception have also been analysed with respect to the CENELEC Standards [26, 27]. EN 50159 states that, when a safety-related electronic system involves the transfer of information between different locations, the transmission system forms an integral part of the safety-related system and it must be proved that the end-to-end communication is safe. Moreover, for the sake of the safety analysis, it is normally useful to consider part of the sender and receiver functionality as belonging to the non-trusted transmission channel. For example, in Level 2, ETCS functions associated with the Eurobalise Transmission System and the Euroradio Transmission System can be thought of as composed of non-trusted and trusted communication parts, see Fig. 6.29.

Figure 6.29 outlines:

- The split of the Eurobalise Transmission System functions into the trusted part, i.e. on-board BTM (trusted), and non-trusted parts, i.e. on-board BTM (non-trusted) and EUB (i.e. Eurobalise);
- The split of the Euroradio Transmission System functions into the trusted part, i.e. on-board EUR (trusted) and Trackside EUR (trusted), and non-trusted parts, i.e. on-board EUR (non-trusted) and Trackside EUR (non-trusted).

Taking into account the several different transmission-related events that can occur in the non-trusted parts, each TRANS-event-n has been further analysed with respect

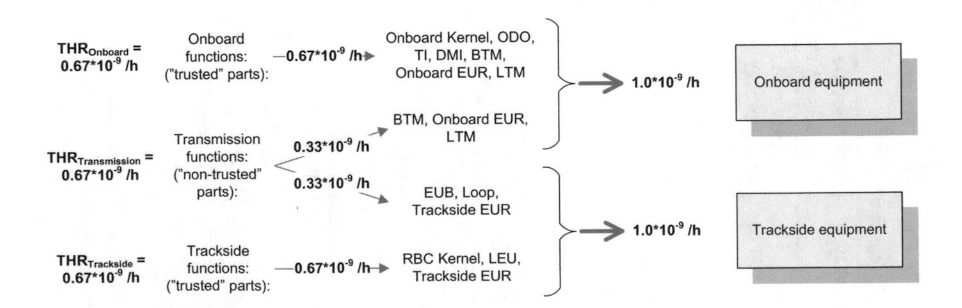

Fig. 6.29 High-level apportionment of top-level ETCS THRs in level 2 to ETCS equipment

to such events; each of these events associated with the non-trusted part belongs to exactly one constituent and one functional element within that constituent. In Fig. 6.30, EUB-Hn and BTM-Hn represent the non-trusted transmission events of the Eurobalise Transmission System, whereas TR-EUR-Hm and OB-EUR-Hm represent the non-trusted transmission events of the Euroradio Transmission System.

As far as the TRANS-xxRADIO n events are concerned, the relationship with the transmission-related events is as follows:

- TRANS-OB/RADIO-1 (Corruption) or TRANS-TS/RADIO-1 (Corruption):
  - TR-EUR-H4: Radio message corrupted in the trackside Euroradio, such that the message appears as consistent.
  - OB-EUR-H4: Radio message corrupted in the on-board Euroradio, such that the message appears as consistent.

  The occurrence of this event is considered to be negligible with respect to THR$_{\mathbf{Transmission}}$ $= 0.67 \times 10^{-9}$ /h due to the characteristics of the safety code[15] [10], and thus, its maximum THR can be set to 0.5×10^{-11}/h assuming an equal apportionment between on-board and trackside. The TRANS-xx/RADIO-1 allocation reflects the bidirectional nature of the radio link and that the potential for corruption is present in either direction (Trackside to On-board or On-board to Trackside). TRANS-OB/RADIO-1 or TRANS-TS/RADIO-1 also includes the undetectable corruption of a message in the air gap.
- TRANS-OB/RADIO-2 (Deletion):
  - TR-EUR-H1: Radio message deleted in the trackside in an undetectable way.
  - OB-EUR-H1: Radio message deleted in the on-board in an undetectable way.

  This event is not classified as a hazard because critical messages sent via RBC are subject to acknowledgement rendering the occurrence of this event negligible.
- TRANS-OB/RADIO-3 (Insertion) or TRANS-TS/RADIO-3 (Insertion) are not considered as hazardous event owing to the measures required by CENELEC EN 50159 [26] such as message sequencing, time stamping and addressing.

Therefore, the total contribution of the Radio Transmission System to the ETCS Core Hazard is negligible, and thus, **THR$_{\mathbf{Transmission}}$ $= 0.67 \times 10^{-9}$ /h can be allocated to the Eurobalise Transmission System only** (see Fig. 6.31).

With regard to the TRANS-BALISE n events [23], the relationship with the transmission-related events is reported below:

- TRANS-BALISE-1 (Corruption):
  - BTM-H4: Transmission to the on-board Kernel of an erroneous telegram, interpretable as correct, due to failure within the on-board BTM function.
  - EUB-H4: Transmission of an erroneous telegram, interpretable as correct, due to failure within a Balise.

[15] This is a design commitment that must be guaranteed and verified.

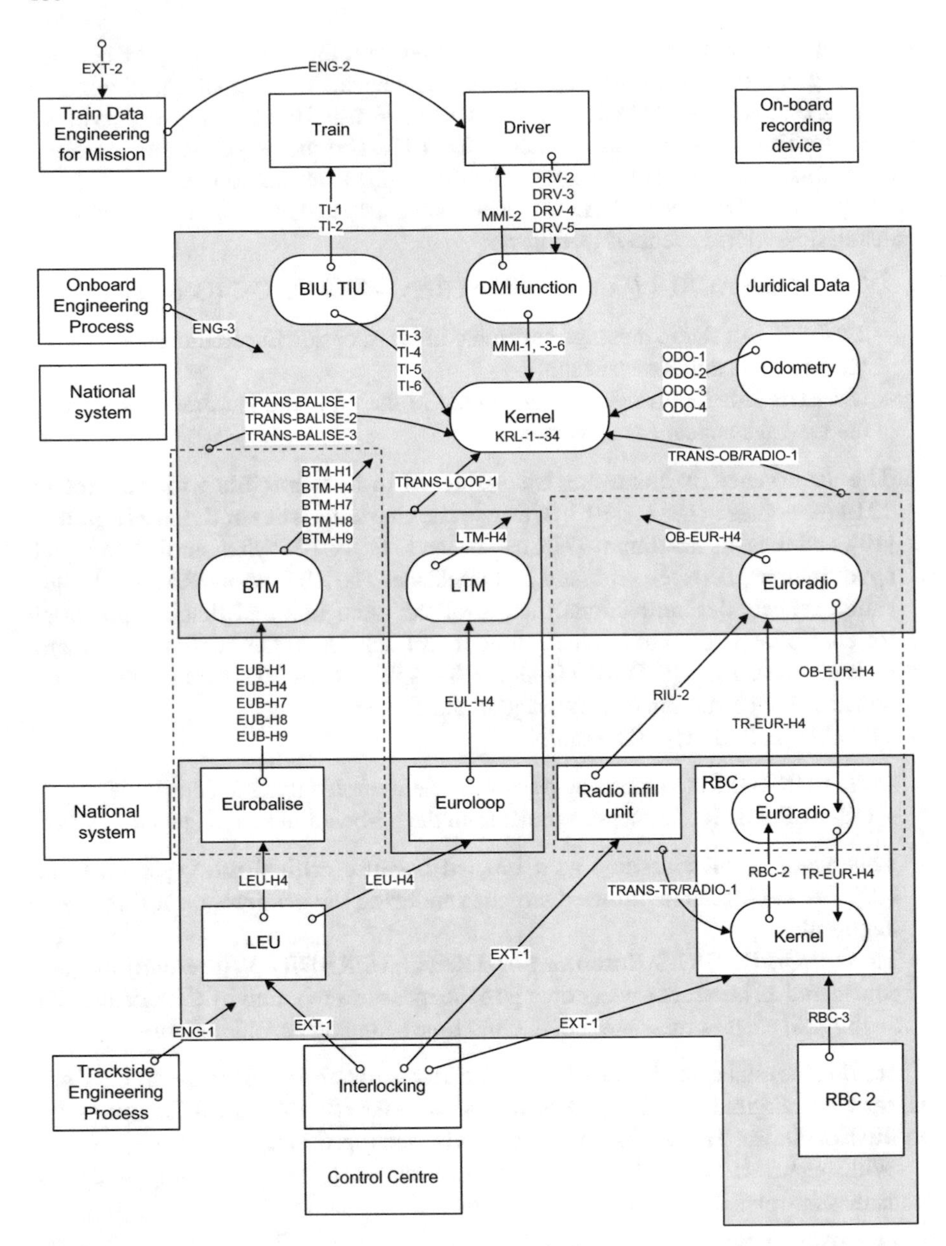

Fig. 6.30 Graphical representation of ETCS hazardous and transmission events

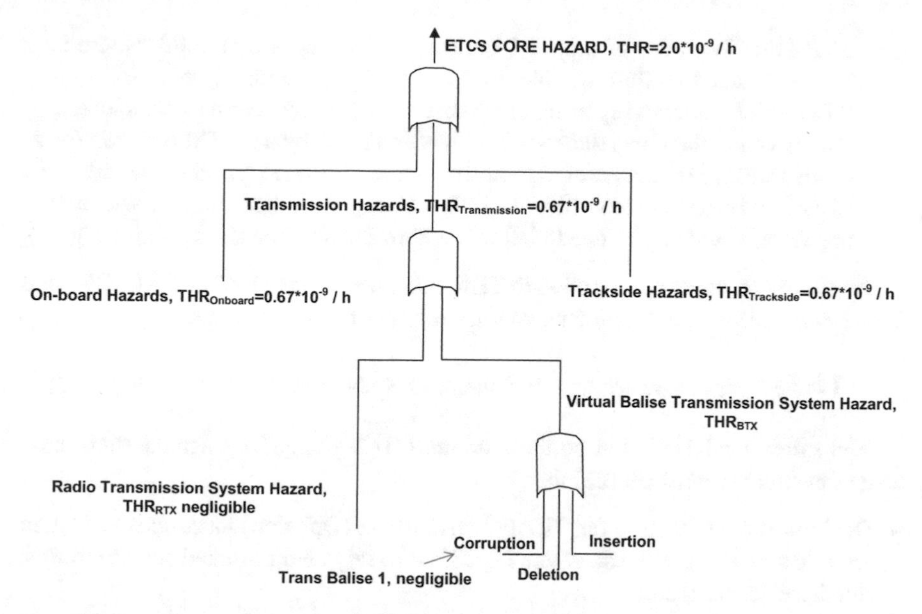

Fig. 6.31 ETCS core hazard apportionment among on-board, transmission and trackside hazards

The occurrence of this event is considered to be negligible with respect to THR$_{\mathbf{BTX}}$ $\approx 0.67 \times 10^{-9}$**/h due to the characteristics of the safety code**[16] [9], and thus, its maximum THR is 0.5×10^{-11}/h assuming an equal apportionment between on-board and trackside. TRANS-BALISE-1 also includes the undetectable corruption of a message in the air gap.

- TRANS-BALISE-2 (Deletion):
  - BTM-H1: A Balise Group is not detected, due to failure within the on-board BTM function.
  - EUB-H1: A Balise Group is not detected, due to failure of the Balise Group to transmit a detectable signal.
- TRANS-BALISE-3 (Insertion):
  - BTM-H7: Erroneous localization of a Balise Group, with reception of valid telegrams, due to failure within the on-board BTM function (erroneous threshold function or significantly excessive tele-powering signal).
  - EUB-H7: Erroneous localization of a Balise Group, with the reception of valid telegrams, due to failure within Balises (too strong uplink signal).
  - BTM-H8: The order of reported Balise, with reception of valid telegrams, is erroneous due to failure within the on-board BTM function (erroneous threshold function or significantly excessive tele-powering signal.

[16]This is a design commitment that must be guaranteed and verified.

- EUB-H8: The order of reported Balises, with reception of valid telegram, is erroneous due to failure within a Balise (too strong uplink signal).
- BTM-H9: Erroneous reporting of a Balise Group in a different track, with reception of valid telegrams, due to failure within the on-board BTM function (erroneous threshold function or significantly excessive tele-powering signal).
- EUB-H9: Erroneous reporting of a Balise Group in a different track, with reception of valid telegrams, due to failures within Balises (too strong uplink signal).

Assuming an equal contribution to THR $_{Transmission}$ from TRANS-BALISE-2 and TRANS-BALISE-3, the following condition has to be demonstrated:

$$THR_{TRANS-BALISE-2} = THR_{TRANS-BALISE-3} < 3.3 \times 10^{-10}/\mathrm{h} \qquad (6.3)$$

The safety analysis based on the inherent ETCS protections against these hazardous events has outlined [23] that:

- The high risk of creating the TRANS-BALISE-2 (Deletion) hazardous event can occur in signalling scenario where ETCS has to rely on non-linked Balise Groups, and these scenario are:
  - operational moves prior to the establishment of linking (e.g. entry into an ETCS area from an unfitted area or train movement in Staff Responsible ERTMS Mode);
  - non-linked Balise Groups in a generally linked network (e.g. Temporary Speed Restrictions provided via Eurobalises instead of RBC, Balise Groups that mark the boundaries of Shunting areas);
  - moves which negate linking (e.g. when a train makes a change in the train travel direction, i.e. enter into REVERSING Mode, in an area with linked balises, the on-board ETCS deletes the on-board stored linked information, MA, SSP and Gradient; a new SOM is required for going into the Full Supervision Mode after the detection of the initial and non-linked Balise Group).
- The high risk of creating the TRANS-BALISE-3 hazardous event can occur at Start of Mission and at Entry into ETCS areas.

Therefore, based on the above identified signalling scenarios, signalling operational considerations for these TRANS-BALISE-x hazardous events have been performed under (*a*) the highest conservative assumption; i.e., the occurrence of each event leads inevitably to the ETCS Core Hazard and (*b*) the mission profile defined above in Sect. 6.3.5.

In particular, as far as the TRANS-BALISE-2 hazardous event is concerned, the following results have been obtained:

- The operational moves prior to the establishment of linking lead to the rate of occurrence of the hazard TRANS-BALISE-2, deletion of a Balise Group when linking is not being checked, equal to [22]: $R_{NoLinking}$ = (Rate of meeting information points * Event frequency qualifiers) * ((Probability of Balise Group Failure) + (Probability of the on-board failing to Detect)) where

- Rate of meeting information points = 2 because it is assumed that, in 1-h journey, there are 2 entries into an ETCS area from an unfitted area in the conventional lines;
- Event frequency qualifiers = 1/1000 = 0,001 because it is assumed that, in 1-h journey, 1 in 1000 times the driver tries and passes the unprotected entry signal at danger;
- Probability of Balise Group Failure = Down time associated with not detectable Balise Groups × Rate of occurrence that a Balise Group composed of at least 2 balises[17] fails (i.e. both balises are not detectable) = $T_{\text{Down Time}} \times \lambda_{IP} = 24\,\text{h} \times \lambda_{IP} = 24\ \lambda_{IP}$ where 24 h is the mean down time of a non-detectable Balise Group assumed in the Mission Profile;
- Probability of the on-board failing to detect = On-Board Down time associated with not detectable Balise Groups * Rate of occurrence that BTM does not detect a Balise Group in an unfitted area = $T_{\text{On-Board Down Time}} \times \lambda_{ONB} = 1\,\text{h} \times \lambda_{ONB} = \lambda_{ONB}$ where 1 h is the on-board mean down time of a non-detectable Balise Group in an unfitted area assumed in the Mission Profile.

By substituting the above value in $R_{\text{NoLinking}}$: $R_{\text{NoLinking}} = (2 \times 0.001) \times (24\lambda_{IP} + \lambda_{ONB})$

- In a generally network of linked Balise Groups, the non-linked Balise Groups lead to the rate of occurrence of the hazard TRANS-BALISE-2, deletion of a Balise Group when linking is not being checked, equal to: $R_{\text{NoLinking}}$ = (Rate of meeting information points) * ((Probability of Balise Group Failure) + (Probability of the on-board failing to Detect))
  where

  - Based on an average number of information points passed in a 1-h journey ≈400 and the ratio of linked information points to non-linked ≈1000 to 4 for conventional lines or 1000 to 1 for high-speed lines, the Rate of meeting information points (TSRs via eurobalises) is respectively equal to 400 * 4/1000 = 1.6 for conventional lines and 0.4 for high-speed lines;
  - Probability of Balise Group Failure = 24 λ_{IP}, as above;
  - The exposure time to on-board failures is based on the time between linked balises with a linking reaction. It is assumed that only 10% of the Balise Groups, i.e. 40 BGs, have such a reaction in level 2. Thus, $T_{\text{On-Board Down Time}}$ = 10/400 = 0.025 h;
  - Probability of the on-board failing to Detect = On-Board Down time associated with not detectable Balise Groups * Rate of occurrence that BTM does not detect a Balise Group = $T_{\text{On-Board Down Time}} \times \lambda_{ONB} = 0.025\,\text{h} \times \lambda_{ONB} = 0.025\lambda_{ONB}$.

[17]The analysis assumes that Balise Groups are composed of at least two Balises when they have to transmit safety data.

By substituting the above value in $R_{NoLinking}$:
$R_{NoLinking} = (0.4 \times 1) \times (24\lambda_{IP} + 0.025\lambda_{ONB})$ for high-speed lines,
$R_{NoLinking} = (1.6 \times 1) \times (24\lambda_{IP} + 0.025\lambda_{ONB})$ for conventional lines.

- Moves which negate linking lead to the rate of occurrence of the hazard TRANS-BALISE-2, deletion of a Balise Group when linking is not being checked, equal to: $R_{NoLinking}$ = (Rate of meeting information points) * ((Probability of Balise Group Failure) + (Probability of the on-board failing to Detect))
  where
  - Rate of meeting information points = 2 because it is assumed that, in 1-h journey, there are 2 SoM;
  - Event frequency qualifiers = 1/1000 = 0,001 because it is assumed that, in 1-h journey, 1 in 1000 times the driver tries and passes the unprotected entry signal at danger;
  - probability of Balise Group Failure = $24\lambda_{IP}$, as above;
  - The exposure time to on-board failures is 1 h, as above;
  - Probability of the on-board failing to Detect = On-Board Down time associated with not detectable Balise Groups * Rate of occurrence that BTM does not detect a Balise Group = $T_{On\text{-}Board\ Down\ Time} \times \lambda_{ONB} = 1\,h \times \lambda_{ONB} = \lambda_{ONB}$.

By substituting the above value in $R_{NoLinking}$: $R_{NoLinking} = (2 \times 0.001) \times (24\lambda_{IP} + \lambda_{ONB})$.

Table 6.8 summarizes the worst-case contributions to $THR_{TRANS\text{-}BALISE\text{-}2}$ based on the operational considerations.

Considering the worst case 2×10^{-10}/h = $(24\lambda_{IP} + \lambda_{ONB})$ and an equal split between the two low-level faults BTM-H1 and EUB-H1, the following maximum failure rate are derived:

1×10^{-10} /h = $24\lambda_{IP}$ => $\lambda_{IP} = 4.1 \times 10^{-11}$ /h
1×10^{-10} /h = λ_{ONB} => $\lambda_{ONB} = 0.4 \times 10^{-7}$ /h

To meet λ_{IP}, it has been introduced the following engineering rule: “When reliance is placed on the detection of unlinked Balise Groups for the announcement of Temporary Speed Restrictions, it is required that **two separate Balise Groups** are used each with a minimum of two balises”.

With this engineering rule, the next worst scenario for $THR_{TRANS\text{-}BALISE\text{-}2}$ becomes 1.6×10^{-7} /h = $(24\lambda_{IP} + \lambda_{ONB})$. By using the same approach adopted before, the following results have been obtained:

0.8×10^{-7} /h = $24\lambda_{IP}$ => $\lambda_{IP} = 3.3 \times 10^{-9}$ /h
0.8×10^{-7} /h = λ_{ONB} => $\lambda_{ONB} = 0.8 \times 10^{-7}$ /h

In the ERTMS community, the following two requirements for technical interoperability have finally been agreed [23]: $THR_{BTM\text{-}H1} = 1.0 \times 10^{-7}$ Failures/h $THR_{EUB\text{-}H1} = 1.0 \times 10^{-9}$ Failures/h

Finally, with regard to $THR_{TRANS\text{-}BALISE\text{-}3}$, the conservative following requirement for technical interoperability has been agreed:

$THR_{TRANS\text{-}BALISE\text{-}3} = 1.0 \times 10^{-9}$ Failures/h

Table 6.8 Summary of the worst-case contributions to THRTRANS-BALISE-2

THRTRANS-BALISE-2	Operational considerations	Note
$R_{NoLinking} = 2 \times 0.001 \times (24\lambda_{IP} + \lambda_{ONB})$ $3.3 \times 10^{-10}\,h^{-1} = 2 \times 10^{-3} \times (24\lambda_{IP} + \lambda_{ONB})$ $1.6 \times 10^{-7}\,h^{-1} = 24\lambda_{IP} + \lambda_{ONB}$	Operational moves prior to the establishment of linking	The scenario about the entry into an ETCS area from an unfitted area assumes that unfitted area has no ATP. This is a very conservative scenario
$R_{NoLinking} = 1.6 \times 1 \times (24\lambda_{IP} + 0.025\,\lambda_{ONB})$ $3.3 \times 10^{-10}\,h^{-1} = 1.6 \times (24\lambda_{IP} + 0.025\lambda_{ONB})$ $2 \times 10^{-10}\,h^{-1} = 24\lambda_{IP} + 0.025\lambda_{ONB}$	Non-linked Balise Groups in a generally linked network	The worst case is for conventional lines. This high demanding THR can be avoided in Level 2 by managing TSRs via RBC only
$R_{NoLinking} = 2 \times 0.001 \times (24\lambda_{IP} + \lambda_{ONB})$ $3.3 \times 10^{-10}\,h^{-1} = 2 \times 10^{-3} \times (24\lambda_{IP} + \lambda_{ONB})$ $1.6 \times 10^{-7}\,h^{-1} = 24\lambda_{IP} + \lambda_{ONB}$	Moves which negate linking	

6.4 Assessing the Conformity and/or Suitability for Use of the ERTMS Constituents and Verifying the ERTMS Subsystems

Chapters 3 and 4 of the TSI [14] set out the essential requirements (i.e. Safety, Reliability and Availability, Health, Environmental Protection and Technical Compatibility) and the required basic parameters (e.g. the THR related to ETCS top hazard gate "Exceedance of safe speed or distance as advised to ETCS", train characteristics, performances of the signalling functions, electromagnetic compatibility between a rolling stock and a CCS trackside equipment) that must be met by **interoperability constituents and CCS subsystems.**

The compliance with these requirements and the required parameters must be demonstrated by [14]:

- Assessing the conformity of the interoperability constituents;
- Verifying the subsystems by applying recommended procedures.

In certain cases, some of the essential requirements may be met by national rules because of:

- The use of Class B systems;
- The known open points in the TSI;
- The derogations under Article 9 of Directive 2008/57/EC; some examples of derogations where one or more TSIs may not be applied are reported below:
  - for any project concerning the renewal or upgrading of an existing subsystem where the loading gauge, track gauge, space between the tracks or electrification voltage in these TSIs are not compatible with those of the existing subsystem;
  - for a proposed new subsystem or for the proposed renewal or upgrading of an existing subsystem in the territory of that Member State when its rail network is separated or isolated by the sea or separated as a result of special geographical conditions from the rail network of the rest of the Community;
  - following an accident or a natural disaster, where the conditions for the rapid restoration of the network do not economically or technically allow for partial or total application of the relevant TSIs;
- The specific cases well described in the TSI applicable to Belgium, UK, France, Poland, Lithuania/Latvia/Estonia, Sweden, Luxembourg and Germany.

In such cases where national rules must be applied, assessment of conformity with those national rules must be carried out under the responsibility of the Member States and by using notified procedures.

The **assessment of an interoperability constituent and/or groups of interoperability constituents** must be carried out by the interoperability constituent manufacturer and by the Notified Body[18] involved by this manufacturer. The manufacturer

[18] A Notified Body is an organization designated by an EU country to assess the conformity of certain products before being placed on the market. These bodies carry out tasks related to conformity

is responsible for providing an **"EC" declaration of conformity** in accordance with Article 13(1) and Annex IV to Directive 2008/57/EC by using one of the following foreseen modules (described in detail in the Commission Decision 2010/713/EU):

- Either the type-examination procedure (Module CB) for the design and development phase in combination with the production quality management system procedure (Module CD) for the production phase; or
- The type-examination procedure (Module CB) for the design and development phase in combination with the product verification procedure (Module CF); or
- The full quality management system with design examination procedure (Module CH1).

Independently of the selected module, the TSI recommends a list of activities (see TSI, Table 6.1) to be carried out during the assessment process of an interoperability constituent and/or groups of interoperability constituents. Particular attention must be given to assess the conformity of the on-board ETCS interoperability constituent owing to its complexity and its critical role in achieving interoperability. In particular, the Notified Body must check that a representative sample of the interoperability constituent has been submitted to a full set of test sequences including all test cases necessary to check the on-board ETCS functions. The applicant (i.e. the manufacturer) is responsible for defining the test cases and their organization in sequences. These verification tests must be carried out in a laboratory accredited in accordance with Regulation (EC) No 765/2008 of the European Parliament and of the Council[19] with the use of the test architecture and the procedures as specified in the TSI. The laboratory must deliver a full report clearly indicating the results of the test cases and sequences used. On the other end, the Notified Body is responsible for assessing the suitability of test cases and sequences to check compliance with all relevant requirements and to evaluate the results of tests in view of the certification of the interoperability constituent. To increase the confidence that the On-board ETCS Interoperability Constituent will operate correctly when installed in CCS on-board subsystems running on different CCS Trackside Subsystems, it is recommended that On-board ETCS Interoperability Constituent is also tested using relevant operational test scenarios from the ones published by ERA. An "operational test scenario" means the description of the intended railway system operation in situations relevant for ERTMS (e.g. entry of a train into an equipped area, awakening of a train, overriding a signal at stop), by means of a sequence of trackside and on-board events related to or influencing the interoperable constituent (e.g. sending/receiving messages, exceeding a speed limit, actions of ERTMS operators, the specified timing between them). The operational test scenarios are based on the engineering rules adopted for the assessment. The tests can be performed using real equipment or a simulated CCS Trackside Subsystem. These additional tests (i.e. those obtained from

assessment procedures set out in the applicable legislation, when a third party is required. The European Commission publishes a list of such notified bodies [Decision 768/2008/EC].

[19]Regulation (EC) No 765/2008 of the European Parliament and of the Council of 9 July 2008 setting out the requirements for accreditation and market surveillance relating to the marketing of products and repealing Regulation (EEC) No 339/93 (OJ L 218, 13.8.2008, p. 30).

the ERA Test suite) are not mandatory for the certification of the On-board ETCS Interoperability Constituent. The applicant[20] for certification of the interoperability constituent may decide to perform them and have them assessed by a Notified Body; the corresponding documentation must provide information about the operational test scenarios against which the interoperability constituent has been checked and whether tests have been carried out with simulators or using real equipment, including type and version of such equipment. Performing these tests at the level of interoperability constituent may also reduce the amount of verification checks at the level of Control-Command and Signalling Subsystem.

The "EC" Declaration of conformity at any interoperability constituent level must report (*a*) which optional and additional functions the interoperable constituent implements and (*b*) the applicable environmental conditions.

As far as the assessment procedures for **CCS Subsystems** are concerned, the applicant's request must also include the "EC" verification of the related CCS subsystem. This **"EC" verification** must be carried out by the involved Notified Body and must be prepared in accordance with Annex VI to Directive 2008/57/EC. On the other side, the applicant must draw up the **'EC' declaration of verification** for the CCS subsystem in accordance with Article 18(1) and Annex V of Directive 2008/57/EC and by using one of the following modules as specified in the TSI:

(*1*) For verifying the CCS On-board Subsystem, the applicant may choose either:

- The type-examination procedure **(Module SB)** for the design and development phase in combination with the production quality management system procedure **(Module SD)** for the production phase; or
- The type-examination procedure **(Module SB)** for the design and development phase in combination with the product verification procedure **(Module SF)**; or
- The full quality management system with design examination procedure **(Module SH1)**.

(*2*) For verifying the CCS Trackside Subsystem, the applicant may choose either:

- The unit verification procedure **(Module SG)**; or
- The type-examination procedure **(Module SB)** for the design and development phase in combination with the production quality management system procedure **(Module SD)** for the production phase; or
- The type-examination procedure **(Module SB)** for the design and development phase in combination with the product verification procedure **(Module SF)**; or the full quality management system with design examination procedure **(Module SH1)**.

---

[20]In accordance with the DIRECTIVE 2012/34/EU OF THE EUROPEAN PARLIAMENT AND OF THE COUNCIL of 21 November 2012 establishing a Single European Railway Area (recast), an applicant means a Railway Undertaking or an international grouping of Railway Undertakings or other persons or legal entities, such as competent authorities under Regulation (EC) No 1370/2007 and shippers, freight forwarders and combined transport operators, with a public-service or commercial interest in procuring infrastructure capacity.

Tables 6.2 and 6.3 of the TSI [14] show the verification checks that must be carried out when respectively verifying a CCS On- board Subsystem and a CCS Trackside Subsystem. In particular and independently of the module chosen, (*a*) the verification of a CCS On-board Subsystem aims at demonstrating that it complies with basic parameters when it is integrated into the vehicle and (*b*) the functionality and performances of interoperability constituents already covered by their EC Declaration of conformity do not require additional verifications. Moreover, the verification of a CCS Trackside Subsystem requires that at least the following application-specific information are provided:

- Line characteristics such as gradients, distances, positions of route elements and Eurobalises/Euroloops, locations to be protected, etc.;
- The signalling data and rules to be handled by the ETCS system.

Member States ensure that, when the process of EC verification of a CCS Trackside Subsystem is initiated, the engineering rules and the preliminary operational test scenarios related to the interactions of its ERTMS parts with the corresponding parts of a CCS on-board subsystem are soon made available to the European Railway Agency. The European Railway Agency must be informed of any changes to operational tests scenarios used during the EC verification. The set of engineering rules for the trackside parts of ERTMS and related operational test scenarios for the CCS Trackside Subsystem must be made available and must be sufficient to describe all intended system operations in normal and identified degraded situations. On the other side, the European Railway Agency (*a*) publishes the engineering rules for the trackside parts of ERTMS and the operational test scenarios and (*b*) collects and properly manages all the comments/clarifications it might receive.

When an **interoperability constituent** does not implement all functions, performance and interfaces specified in the TSI, an EC certificate of conformity may only be issued if the unimplemented functions, interfaces or performance is not required to integrate the interoperability constituent into a subsystem for the use indicated by the applicant.

On the other end, if a CCS **subsystem** does not implement all functions, performance and interfaces of the TSI (e.g. because they are not implemented by an interoperability constituent integrated into it), the certificate of verification must indicate which requirements have been assessed and the corresponding conditions and restrictions on the use of the subsystem and its compatibility with other subsystems. Therefore, interoperability constituents and CCS subsystems that do not implement all functions, performance and interfaces as specified in the TSI can obtain EC certificates of conformity or, respectively, certificates of verification, under the following conditions:

- The applicant for EC verification of a CCS Trackside Subsystem is responsible for deciding which functions, performance and interfaces need to be implemented to meet the objectives for the service and to ensure that no requirements contradicting or exceeding the TSIs are exported to the on-board Control-Command and Signalling Subsystems;

Table 6.9 Mandatory standards to be applied in the certification process.

Reference	Document name and comments	Version
EN 50126	Railway applications The specification and demonstration of reliability, availability, maintainability and safety (RAMS)	1999
EN 50128	Railway applications Communication, signalling and processing systems Software for railway control and protection systems	2001 or 2011
EN 50129	Railway applications Communication, signalling and processing systems Safety-related electronic systems for signalling	2003
EN 50159	Railway applications Communication, signalling and processing systems Safety-related communication in transmission systems	2010

- The operation of a CCS on-board subsystem that does not implement all functions, performance and interfaces specified in the TSI may be subject to conditions or restrictions due to compatibility and/or safe integration with CCS Trackside Subsystems. The applicant for EC verification is responsible for ensuring that documentation[21] accompanying the declaration provides all the information that an operator needs to identify such conditions and restrictions;

The Member State may refuse for duly justified reasons the authorization for placing in service, or place conditions and restrictions on the operation, of CCS subsystems that do not implement all functions, performance and interfaces specified in this TSI.

Table 6.9 lists the mandatory standards to be applied in the certification process.

6.5 Market and Cost Constraints

In the context of ERTMS framework, the ERTMS Community has reached the following important results:

- Stable specifications: ERTMS SRS 3.6.0 is functionally complete. The Railway Interoperability & Safety Committee (RISC) has adopted them in June 2016 [28];
- 4th Railway Package: ERA will get new powers for trackside approval and vehicle authorization - Powers commence June 2019;
- ERTMS Deployment Plan (EDP) adopted [7]:

[21] Directive 2008/57/EC [11] reports the minimum contents of the required documentation (named Technical File).

  - 7200-km ERTMS lines operational by 2019, including crucial cross-border sections;
  - Advanced deployment plans in several Member States;

- 2016 Memorandum of Understanding (MoU) for supporting a collective and a disciplined approach,

However, the European IMs and the European Railway Undertakings (RUs) have asked for cost-effective ERTMS solutions suitable for all the different types of railway lines and not for the high-speed lines only. In particular, they are pushing for ERTMS solutions suitable for local and regional lines where capital investments are limited. In addition, the European Railways Agency has recognized the need of still innovating ERTMS taking into account the following key pillars:

- Stability (RUs/IMs/Suppliers) of the Specifications (backward compatibility) to protect existing investments (trackside/on-board);
- Cost reduction of ERTMS system:
  - Trackside: backwards compatibility in case of new functions;
  - On-board: no new mandatory functions on top of BL3 R2 during long period;
- Stability supported by fast error corrections process for existing functions;
- Need for evolution in ERTMS specifications to increase line capacity.

In particular, ERA has set up a Technical and a Financial Frameworks with the following objectives:

- Technical Framework:
  - Coordination between R&D Programs and the ERA program (i.e. Joint, clear communication and dissemination among all actors);
  - Managing risk of increasing complexity of ERTMS system (i.e. Harmonized operation concept in nominal and degraded modes, standardized engineering rules);
  - Helping early adopters by the validation of specifications process before legal release of new set specifications (i.e. support of structured approaches; simulation tools; operational feedback from early implementers).
- Financial Framework:
  - To support the evolution of ERTMS specifications;
  - To ensure a fair allocation of benefits, costs and risks between IMs and RUs in case the deployment of a new function/new release improves the overall business case.

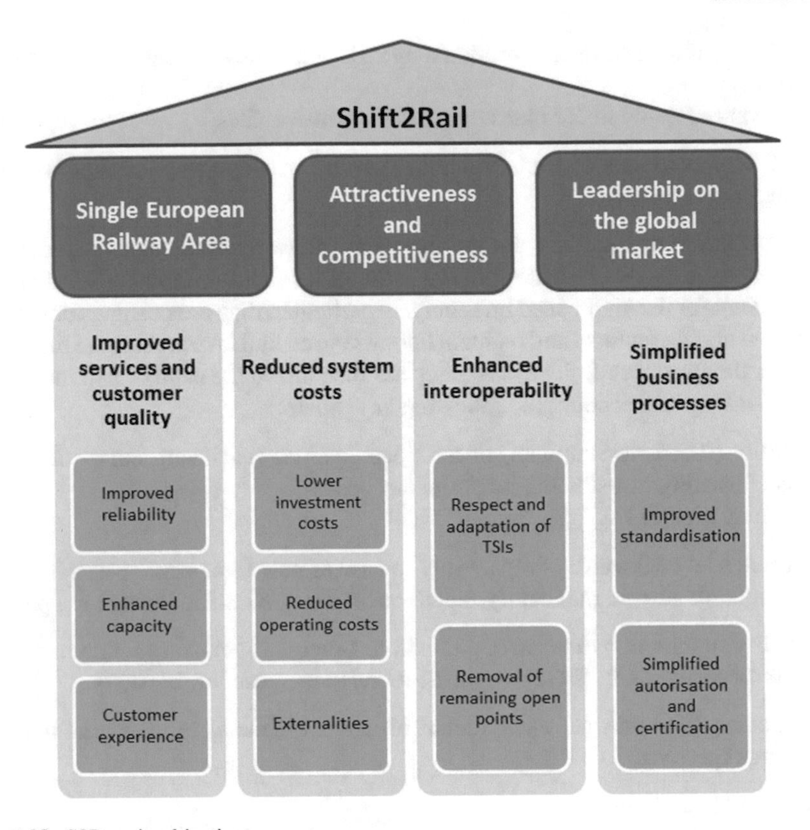

Fig. 6.32 S2R main objectives

Moreover, the Council Regulation (EU) No 642/2014 of 16 June 2014 (Shift2Rail (S2R) Regulation) has approved the important S2R programme having the following main objectives, see Fig. 6.32:

- To create a Single European Railway Area;
- To improve the Attractiveness & Competitiveness of the European Railway System;
- To gain and to keep the Leadership on the global market.

In particular, in the context of the S2R IP2 programme, the Technology Demonstrator named Technology Demonstrator (TD) 2.4—Fail-Safe Train Positioning (including satellite technology) has been defined with the main objectives of carrying out the:

Definition, development and verification of a Fail-Safe Train Positioning (including satellite technology) as a component of the ERTMS/ETCS.

This solution will be based on a safe on-board multisensor positioning functional block, where GNSS is the preferred technology. The key properties of such a solution will be:

- Standard and interoperable solution (expected to be part of a new ERTMS Baseline);
- Based on the state-of-the-art technologies in the use of:
  - Absolute position technologies (e.g. GNSS and the Augmentation Subsystem) starting from the main results reached in the related GSA projects (e.g. ERSAT-EAV [4], STARS [6], RHINOS [5]) and the ESA projects(e.g. 3InSat [2] and SBS [3]).
  - Kinematic sensor technologies (e.g. IMU sensors).

TD2.4 will be implemented by means of two consecutive R&D projects, respectively named X2Rail-2 whose start date is September 2017 with a duration of 3 years and X2Rail-5 with completion date by the end of 2022. In the context of all these innovative R&D initiatives, the evolution of ERTMS is based on the replacement of the major part of Physical Balises with Virtual Balises (see next chapter) and on the enhancement of the odometry performance. Both actions have the objectives of reducing the capital expenditures (CAPEX) costs (i.e. costs for the procurements and the installations of the physical balises) and the operating expenses (OPEX) costs (i.e. the operational costs associated with the maintenance of the Balise Groups and/or the replacement of faulty physical balises).

6.6 RF Channel Impairments on Trains

The train environment is a very critical and high demanding environment with respect to the typical classes of GNSS radio frequency impairments [29] such as Multipath (described in Sect. 2.1.5.1), RFIs (described in Sect. 2.1.5.2), (NLOS) conditions (described in Sects. 4.4 and 5.4.6).

In the railway environment, the GNSS signal that reaches the on-board unit is affected by different types of multipath as it travels at the same level as obstacles present on the railway line such as trees, bushes, other trains, nearby buildings and urban canyons. Figure 6.33 shows some train environment conditions where critical multipath phenomena can occur. On the other hand, Fig. 6.34 outlines some locations where unintentional RFIs might occur.

In general, contrary to what has been specified, designed and validated for avionics applications, the effects of these feared events cannot be considered negligible any more. Feared events from both the system and the local environments contribute to positioning errors that can lead to an increased risk of hazardous misleading information (i.e. position not bounded by confidence interval). It is particularly difficult to detect, yet alone correctly estimate the multipath delay (for example) by simply examining and processing GNSS signals or navigation and observation (RINEX like) data in real time. Moreover, the current GNSS avionic models do not seem to adequately represent multipath in the railway environment.

The figures below outline the results of some GNSS SIS measurements carried out by using a COTS GNSS receiver installed on-board a train travelling amid a

Fig. 6.33 Examples of hostile train environments (Multipath)

Fig. 6.34 Examples of hostile train environments (RFI)

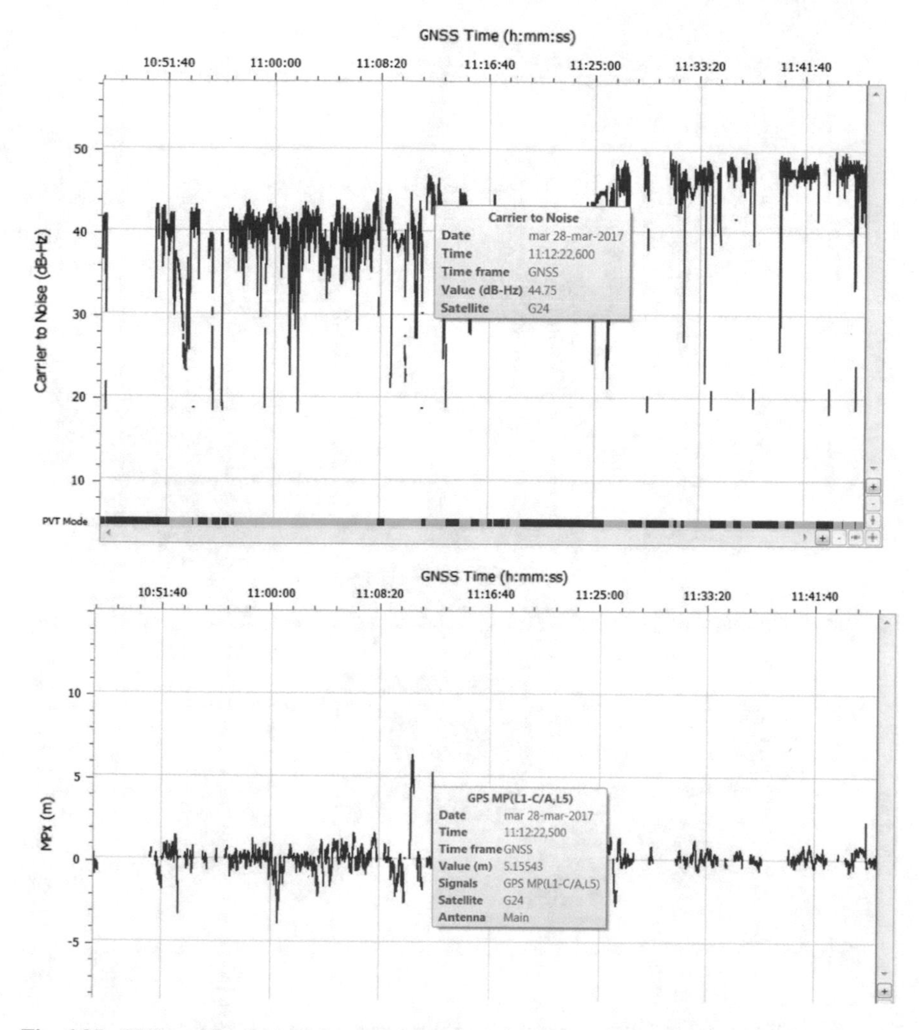

Fig. 6.35 GPS satellite G24 carrier-to-noise ratio and relative multipath estimate

series of tunnels on 28 March 2017 from La Spezia to Parma, two Italian Railway Stations of a Regional Line.

Figure 6.35 represents the multipath estimated using the CMC method and double-frequency combinations of carrier phase measurements on GPS L1 C/A and L5 frequencies for the GPS satellite G24 having a C/No value close to 45 dB-Hz and an elevation of 25° (not particularly low). **The computed multipath is bounded to 6 m whereas the multipath computed in accordance with the WAAS MOPS (Appendix J.2.4 of** [30]**) and LAAS MOPS (Sect. 2.3.12.1 of** [31]**) is underestimated.**

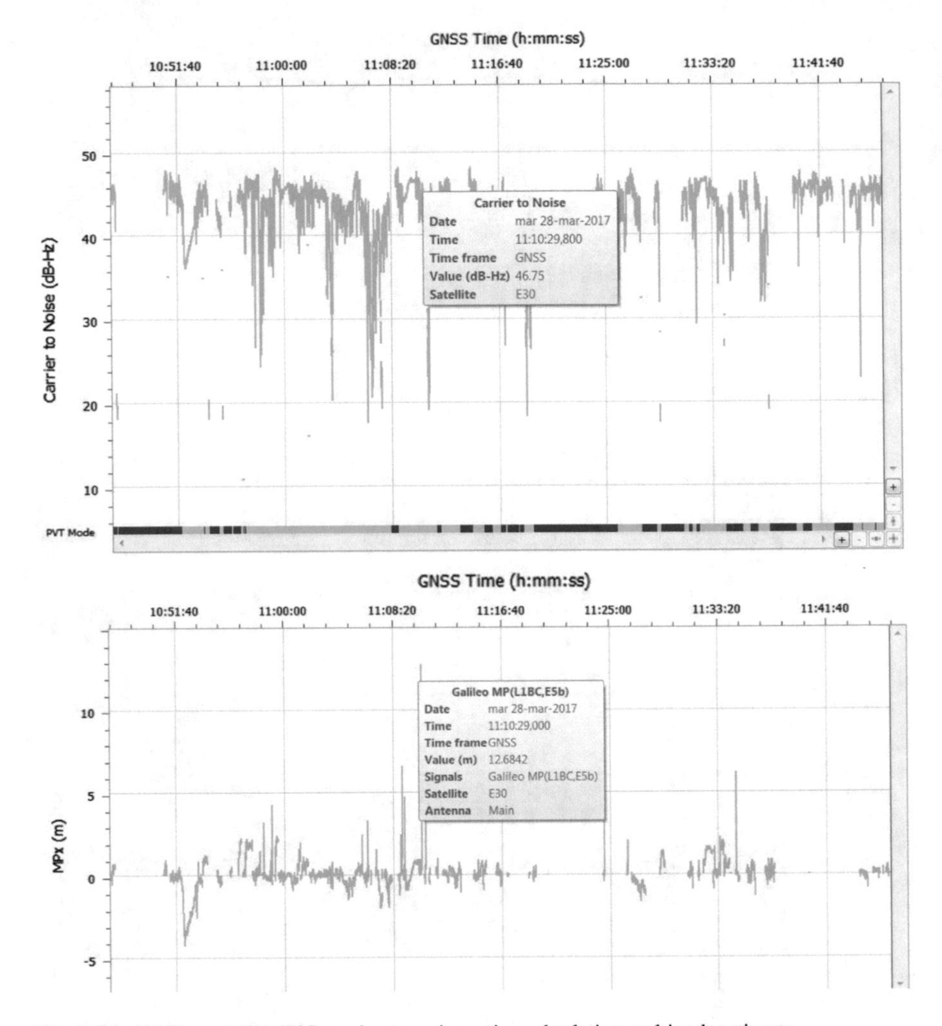

Fig. 6.36 Galileo satellite E30 carrier-to-noise ratio and relative multipath estimate

Figure 6.36 showcases a situation where the Galileo satellite E30 signal reception at post-correlation results in a C/N0 value near to 47 dB-Hz (Galileo satellite with an elevation of 35°) and the **computed multipath is bounded to 13 m whereas the multipath computed in accordance with the WAAS MOPS (Appendix J.2.4 of** [30]**) and LAAS MOPS (Sect. 2.3.12.1 of** [31]**) is underestimated.**

Many field measurement campaigns have been performed in Italy and in Australia in the context of different Ansaldo STS projects, and they have confirmed the peculiarities of the railway environment.

The next chapter will describe a possible application of the GNSS Technology and the tailoring of the concept of integrity, described in the first part of the book, that enables the enhancement of the ERTMS signallling system so as to still guarantee the SIL 4 Train Position function in the high demanding rail environment.

References

1. UNISIG. SUBSET-023 v.3.3.0, Glossary of terms and abbreviations
2. 3INSAT project (2016) Train integrated safety satellite system. https://business.esa.int/projects/3insat
3. SBS-Rails project, Space based services for railway signalling. https://business.esa.int/projects/sbs-rails
4. ERSAT-EAV project (2017) ERTMS on SATELLITE enabling application & validation. http://www.ersat-eav.eu/
5. RHINOS project, Railway high integrity navigation overlay system. https://www.gsa.europa.eu/railway-high-integrity-navigation-overlay-system
6. STARS project, Satellite technology for advanced railway signalling. http://www.stars-rail.eu/
7. UNISIG. UBSET-026 v3.6.0 ERTMS/ETCS system requirements specification
8. UNISIG. SUBSET-100 v.2.0.0, Interface “G” specification
9. UNISIG. SUBSET-036 v.3.1.0 (2016) FFFIS for Eurobalise
10. UNISIG. SUBSET-037 v.3.1.0, EuroRadio FIS
11. Directive 2008/57/EC of the European Parliament and of the Council of the European Union of 17 June 2008 on the interoperability of the rail system within the community (recast) (text with EEA relevance). http://eur-lex.europa.eu/legal-content/EN/TXT/?uri=celex:32008L0057, 17 June 2008
12. ERA (2015) List of class B systems, ERA/TD/2011-11 V3.0
13. Council of the European Union (1985) Council regulation (EEC) No 2137/85 of 25 July 1985 on the European Economic Interest Grouping (EEIG), 25 July 1985
14. EU. Commission regulation (EU) 2016/919 of 27 May 2016 on the technical specification for interoperability relating to the control-command and signalling subsystems of the rail system in the European Union, 27 May 2016
15. ERA (2016) Guide for the application of the TSI for the subsystems control-command and signalling track-side and on-board “ERA/GUI/07-2011/INT V4.0”
16. Decision No 661/2010/EU of the European Parliament and of the council of 7 July 2010 on Union guidelines for the development of the trans-European transport network - (recast), 7 July 2010
17. Commission implementing regulation (EU) 2017/6 of 5 January 2017 on the European rail traffic management system European deployment plan, 5 Jan 2017
18. ERA (2015) Report on ERTMS longer term perspective V1.5
19. UNISIG. SUBSET-040 v.3.4.0, Dimensioning and engineering rules
20. UNISIG. SUBSET-041 v.3.2.0, Performance requirements for interoperability
21. UNISIG. SUBSET-088 Part 0 v.3.6.0, ETCS application levels 1&2 safety analysis: part 0 - document overview
22. CENELEC. EN 50126 Railway applications - The specification and demonstration of reliability, availability, maintainability and safety (RAMS)
23. UNISIG. SUBSET-088 Part 3 v.3.6.0, ETCS application levels 1&2 safety analysis: part 3 THR apportionment
24. ERTMS Users group (1998) ERTMS/ETCS RAMS Requirements specification chapter 2 RAM, Reference EEIG: 96S126 V6
25. UNISIG. SUBSET-088-2 Part 2 v.3.6.0, ETCS application level 2 safety analysis: part 2 functional analysis
26. CENELEC. EN 50159 Railway applications - communication, signalling and processing systems - safety-related communication in transmission systems

27. CENELEC (2003) EN 50129 Railway applications communication, signalling and processing systems safety related electronic systems for signalling
28. UNISIG. SUBSET-088-2 Part 1 v.3.6.0, ETCS application level 2 safety analysis: part 1 functional fault tree
29. Kaplan E, Hegarty C (2006) Undestanding GPS: principles and applications, 2nd edn. Artech House
30. DC Radio Technical Commission for Aeronautics Special Committee 159 SC-159, Washington (2006) Minimum operational performance standards for global positioning system/wide area augmentation system airborne equipment. RTCA DO-229D
31. DC Radio Technical Commission for Aeronautics Special Committee 159 SC-159, Washington (2008) Minimum Operational Performance Standards (MOPS) for GPS Local Area Augmentation System (LAAS) Airborne Equipment. RTCA DO-253C

Chapter 7
A New Perspective for GNSS Based Safe Train Position

Salvatore Sabina, Nazelie Kassabian and Fabio Poli

Abstract This chapter describes the virtual balise concept and summarizes the benefits associated with its use in the evolution of the ERTMS system. A possible-enhanced ERTMS functional architecture suitable for implementing the virtual balise concept is also presented along with the detailed description of the main new functional blocks. This chapter also introduces the proposed extensions of the key ERTMS concepts for estimating the train position based on virtual balises, and consequently, an innovative railway integrity concept is described based on the peculiarities of the railway operational rules and the signalling principles for guaranteeing safe train movements. Finally, a preliminary apportionment of the ETCS Core Hazard tolerable hazard rate based on the use of not only physical balises but also virtual balises is presented.

7.1 The Impact of the Use of Virtual Balise on ERTMS/ETCS

The main motivation for the introduction of GNSS positioning technology in ERTMS/ETCS train positioning function is economical (a) to reduce the global costs incurred in terms of CAPEX and OPEX without incurring major impact on the current ETCS implementation, and, thus, (b) to start the process of developing

S. Sabina (✉)
Ansaldo STS S.p.A, Via Paolo Mantovani 3-5, 16151 Genova, Italy
e-mail: salvatore.sabina@ansaldo-sts.com

N. Kassabian
Ansaldo STS S.p.A, Via Volvera 50, 10045 Piossasco, Turin, Italy
e-mail: nazelie.kassabian@ansaldo-sts.com

F. Poli
Ansaldo STS S.p.A, Via Ferrante Imparato 184, 80147 Naples, Italy
e-mail: fabio.poli@ansaldo-sts.com

© Springer International Publishing AG, part of Springer Nature 2018

L. Lo Presti and S. Sabina (eds.), *GNSS for Rail Transportation*, PoliTO Springer Series, https://doi.org/10.1007/978-3-319-79084-8_7

new cost-effective solutions as requested by the marked needs (as summarized in Sect. 6.5). Moreover, in order to reduce the modifications to the existing ERTMS standard, the proposed approach for the introduction of this technology preserves the signalling principles described in Sect. 6.3.4 used to safely determine the Train Position.

The approach adopted by the space and railway communities (under the GSA and ESA guidance [1, 2]) is based on the concept of **Virtual Balise**, i.e. an abstract entity (see Sect. 7.2) that allows the replacement of a large number of physical balises with these abstract entities. Moreover, the proposed solution exploits the intrinsic features of the ERTMS functions and of the available SIL 4 odometry to also guarantee the SIL 4 train positioning in the high-demanding railway environment with respect to the RF channel impairments described in Sect. 6.6. In addition, the use of virtual balises instead of physical balises has also the following indirect benefits:

- High availability of the railways lines because virtual balises do not intrinsically fail (i.e. they are abstract entities); the reduction of the events associated with the failure of the balise group message consistency checks (see Sect. 6.3.4.4) will also have a positive consequence on the schedule adherence and the operational availability performance indexes described in Sect. 6.3.6;
- High availability of the railways lines because any implementation of the virtual balise concept has a more robust noise immunity with respect to physical balises; as for the above bullet, a more robust balise detection function leads to the improvements of the schedule adherence and the operational availability performance indexes;
- High availability of the railways lines with respect to vandalisms (i.e. virtual balises cannot be removed by vandals); the dismount of a physical balise is equivalent to the failure of a physical balise with the activation of the on-board reaction. Therefore, as for the above two bullets, the use of virtual balises would lead to better performances;
- Reset of the train confidence interval with a higher frequency than the corresponding frequency derived by the use of physical balise and, then, a more regular train movements (i.e. reduced number of accelerations/decelerations and applications of brakes).

Due to this consolidated approach that enables the ERTMS/ETCS kernel to apply the same policy independently from the type of balise, backward compatibility is easy to achieve as long as the BTM functions are maintained in the on-board constituent.

7.2 The Virtual Balise Concept

The **physical balise** (i.e. eurobalise) is a physical equipment installed on a sleeper (see Sects. 6.2.5 and 6.3.3). During the design phase of an ERTMS trackside subsystem, the signalling designer establishes the track location where the eurobalise must be installed and the balise telegram or the information that the eurobalise must send to the ERTMS/ETCS on-board platform (with the ERTMS/ETCS kernel as the final destination). The identification of the location and of the information must be done in accordance with the applicable (national/european) signalling rules and the ERTMS/ETCS dimensioning and engineering rules [3]. For a fixed eurobalise (see Sect. 6.2.5), the telegram must be preloaded on the fixed eurobalise before its commissioning. A telegram contains one header and an identified and coherent set of packets [4]. The information part of the balise telegram (also named balise information), i.e. the user bits, is the telegram without CRC, control bits and synchronization bits. The length of the telegram is either 341 bits (including 210 User Bits) which is also referred to as "short telegram", or 1023 bits (including 830 User Bits) which is also referred to as "long telegram".

On the other hand, the **virtual balise** is an **abstract data type** capable of storing the eurobalise user bits associated with a balise telegram. Similar to what the signalling designer does for the physical balise, during the signalling design phase, he/she establishes the track location where such a virtual balise would be **logically** installed (e.g. km 13+212) and the user bits (i.e. the information) that the virtual balise must send to the ERTMS/ETCS kernel **when the estimated GNSS-based position of the GNSS antenna mounted on the train roof and projected on the track (Virtual Antenna reference mark) matches the location established by the signalling designer.**

For example, Fig. 7.1 represents the output of a signalling design phase where the following balise groups have been identified with the related described properties:

- The balise group named BG n is composed of two **physical** balises; the BG **location reference** is the balise B1/2 located **at km 12+120**, the second balise is located at a distance equal to 3 m from the first balise, the balises must be programmed with the user bits identified by the signalling designer;
- The balise group named BG y is made up of two **virtual** balises; the BG **location reference** is the balise VB1/2 located **at km 12+232**, the second balise is located at a distance equal to 3 m from the first balise, the user bits specified by the signalling designer must be respectively stored into the abstract data associated with these virtual balises;

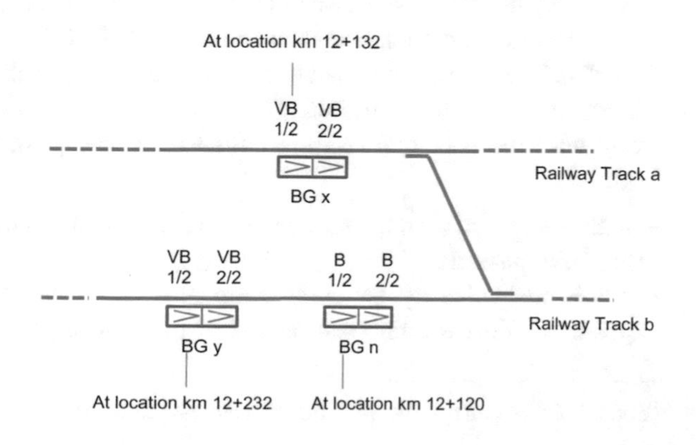

Fig. 7.1 Example of schematic plan with both physical and virtual balises

- The balise group named BG x is composed of two **virtual** balises; the BG **location reference** is the balise VB1/2 located **at km 12+132**, the second balise is located at a distance equal to 3 m from the first balise, the user bits specified by the signalling designer must be respectively stored into the abstract data associated with these virtual balises.

In accordance with this example, two physical balises will be installed on "Railway Track b" with the telegrams associated with the identified user bits. On the other hand, four virtual balises will be stored as abstract data associated with "Railway Track a" and "Railway Track b" in the track database that will be uploaded to the on-board platform during the train mission; each abstract data will include the corresponding user bits and its location in the railway database will be associated with the location identified in the railway schematic plan.

Let us consider a train equipped with (a) an ERTMS/ETCS compliant BTM and related Antenna, (b) a new on-board module named Virtual Balise Reader (VBR) with its related Antenna,[1] and (c) the ERTMS/ETCS kernel.

During the train run, (a) the BTM periodically generates the tele-powering signal to energize any (physical) eurobalise that it can encounter and (b) the VBR periodically computes the estimated GNSS-based position of the GNSS Antenna mounted on the train roof and projected on the track (Virtual Antenna reference mark) and compares it with the locations associated with the virtual balises stored in the on-board track description.

After passing over a physical balise and for each correctly decoded telegram, the BTM provides both the user bits of the decoded telegram and the reference position of the physical balise to the ERTMS/ETCS kernel. On the other hand, when the estimated GNSS position matches the stored position on the on-board track description, VBR provides both the user bits associated with the virtual balise and the reference position of the virtual balise to the ERTMS/ETCS kernel.

Therefore, the ERTMS/ETCS kernel logically receives the same type of information (i.e. user bits and the reference location) independently from the type of medium through which this information is sent: a physical or a virtual balise. The ERTMS/ETCS kernel remains responsible for implementing all the ERTMS/ETCS functions related to balises (e.g. LRBG, linking, expectation window, balise message consistency checks, ...) described in the previous Chap. 6.

For the sake of completeness, as the ERTMS/ETCS kernel must take into account the balise location reference accuracy for periodically estimating the train confidence interval (see Sect. 6.3.4.3), both BTM and VBR must/shall provide such accuracy along with the user bits and the balise reference position. This accuracy is (see Sect. 7.5):

- within +/−1 m with a confidence level as that for SIL 4, when a physical balise has been passed or
- within the value dynamically estimated by the VBR when the match of both positions (GNSS train position and virtual balise position stored in the on-board

[1]The BTM function and the VBR function can be implemented on a unique safe platform.

track description) occurs with an estimated confidence interval based on many factors related to the GNSS SIS.

The integrity associated with the estimated position of the Virtual Antenna reference mark is the key aspect of the virtual balise concept.

7.3 A Possible ERTMS Enhanced Architecture to Integrate the GNSS Positioning Technology

The implementation of the virtual balise concept in ERTMS/ETCS leads to a definition of a possible functional architecture which integrates the GNSS technology in the existing high level ERTMS/ETCS reference architecture by maintaining backwards compatibility. Moreover, the integration of GNSS should be carried out by having in mind a set of principles for virtual balise detection that are in line with railway safety rules, namely EN 50129 [5], ERTMS RAMS requirements specification [6] and EN 50126 [7].

The ERTMS reference functional architecture described in Sect. 6.3 outlines the following three transmission systems, see Fig. 6.12:

- the Eurobalise transmission system,
- the Euroloop transmission system and
- the Radio transmission system.

The proposed ERTMS-enhanced functional architecture foresees the introduction of an additional transmission system, named the virtual balise transmission system (VBTS), see Fig. 7.2.

For the sake of simplicity, Fig. 7.2 depicts those ERTMS/ETCS functions that are relevant for the implementation of the virtual balise concept; all the functions and the interfaces described in the ERTMS/ETCS reference architecture (see Sect. 6.3.3 are still valid and applicable, e.g. those associated with the eurobalise transmission system. The VBTS is a safe-spot transmission-based system, as for the Eurobalise Transmission System, conveying information from the trackside infrastructure in terms of balise information[2] to the on-board equipment. The VBTS main functional blocks are described below.

VBR Functions

The VBR functions can be fulfilled by a chain composed of GNSS antenna, GNSS receiver, a railway-tailored PVT computation and VBD function. The main virtual balise reader functions are:

- To compute the position of the train front-end by using the code and the carrier phase measurements as well as the augmentation and integrity information,

[2]The information part of the Balise Telegram (i.e., the telegram without CRC, control bits, and synchronization bits), i.e., the user bits.

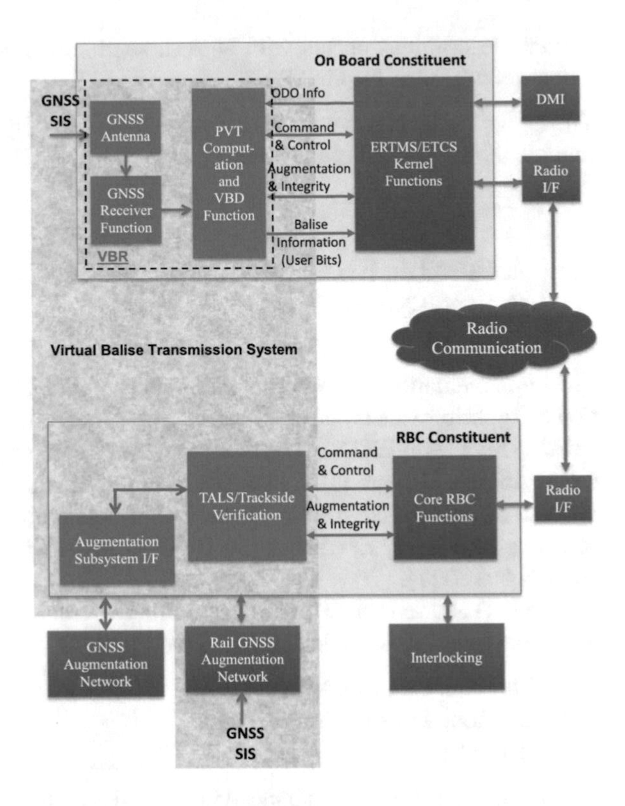

Fig. 7.2 High-level ERTMS/ETCS architecture based on GNSS positioning

- To process augmentation information received from either the Rail GNSS Augmentation Network or the GNSS Augmentation Network and compute the PL associated with the train front-end position to detect virtual balises.
- To detect virtual balises by using the above-computed train front-end position and pre-known virtual balise positions stored into an on-board track description,
- To provide the following information to the ETCS on-board kernel when a balise passage occurs:
  - Time / odometer stamp of the detected virtual balise centre[3] and the dynamically computed virtual balise detection accuracy (i.e. the on-board estimated maximum virtual balise location error computed at the detection time in accordance with the pre-established THR) and
  - User bits of the balise telegram (i.e. the balise information) for the detected virtual balise, stored into an on-board track description,
- To guarantee the delivery of virtual balises in the correct sequence,

[3]It is the instant and/or the location when the Virtual Antenna reference mark, computed by means of the GNSS technology, crosses over the nominal position of the virtual balise in accordance with the track description. The position of the Virtual Antenna is an ideal position below the train where a BTM Antenna would detect ideal physical balises placed in accordance with the track description.

- To guarantee crosstalk protection,
- To ensure immunity to environmental noise,
- To detect user-bit errors,
- To execute start-up tests,
- To dynamically execute run-time tests to detect failures in the virtual balise detection function; this anomaly must be reported to the ERTMS/ETCS Kernel.

ERTMS/ETCS Kernel and Core RBC Functions:

The ERTMS/ETCS kernel is the core of the ERTMS/ETCS on-board equipment [8]. Both the ERTMS/ETCS kernel and the module Core RBC Functions have a double role in the context of VBTS:

(a) the implementation of all the ERTMS/ETCS functions in accordance with [9] and
(b) the gateway function for enabling the cooperation between the (on-board) VBR function and the TALS/trackside verification function via new dedicated packets exchanged with Euroradio messages.

TALS/Trackside Verification Functions:

The track area location server (TALS)/trackside verification functional block is responsible for either:

- computing the GNSS augmentation information based on the information periodically acquired from the Rail Reference Stations of the Rail GNSS Augmentation Network and disseminates it to the on-board; a spatial and temporal validity is associated with each GNSS augmentation information, or
- disseminating the GNSS augmentation information received from the GNSS Augmentation Network to the on-board, and
- based on the position information obtained from the Core RBC Functions functional block and timely warning information received from the augmentation network, computing and disseminating alarms to the on-board so as it can apply specific reactions, when required.

Rail GNSS Augmentation Network Functions:

The Rail GNSS Augmentation Network is a ground-based augmentation network that is made up of cheap georeferenced railway Reference Stations acting as railway peripheral posts. These Reference Stations are compliant with the applicable CENELEC Standard such as EN 50126 [7], EN 50128 [10] and EN 50129 [5]. They are geographically distributed along the track to always guarantee the presence of two independent Reference Stations within the radius of influence for the on-board position (the radius of influence for GNSS correction is several tens of kilometres, e.g. 80 km). Two Rail Reference Stations are considered independent when their relative position guarantees statistical independence of GNSS measurements (distance in the order of magnitude of about a hundred of metres typically allows achieving it).

GNSS Augmentation Network Functions:

The GNSS Augmentation Network is, instead, a public available augmentation system, e.g. SBAS or GBAS-based suitable for providing the Safe-Of-Life service compliant with safety, performance and quality railway requirements. This type of Augmentation Network has been investigated in [11, 12] and further studies are still in progress.

This proposed ERTMS Enhanced architecture foresees that the trackside interoperable RBC component is responsible for disseminating the augmentation information to on-board because many field test measurement campaigns [13, 14] have demonstrated that the quality and the availability of the GNSS SBAS Geostationary SIS measured along many railway lines do not seem adequate for guaranteeing the required ERTMS performance. Therefore, as the radio communication coverage between the trackside RBC platform and the on-board platform is guaranteed for allowing the train mission, the same ERTMS communication session can be used for transferring augmentation information from the RBC platform to the on-board platform. This guarantees the dissemination of the augmentation information for all the train mission. Moreover, as the ERTMS communication session enables safe and secure (in terms of security) exchanges of messages between both RBC and on-board platform, the use of this communication session also guarantees the dissemination of augmentation information in accordance with the CENELEC EN 50159 [15].

Part I of this book has described in detail the important role of an augmentation system as an independent diagnosis of the entire GNSS system (i.e. the combined ground and airborne subsystems), to detect feared events originating from the GNSS system (e.g. ephemeris errors, satellite failures, satellite clock errors, errors induced by the ionosphere and the troposphere, signal deformation). Moreover, Part I of this book has also defined the augmentation information (i.e. augmentation data) as the correction and integrity information to be provided to the on-board (where the end-user GNSS receiver is installed) to enable improvement of positioning accuracy and the computation of a protection level.

On the other hand, the TALS/Trackside Verification function is responsible for carrying out additional railway verification checks on the train position determined via the on-board position report and the on-board applied pseudorange corrections. These additional checks will be based on, for example, coherence checks with the track occupancy, the sequence of occupancy of adjacent tracks, the knowledge of the status of the line or station, etc...

Adopting the same concepts of trusted and non-trusted parts used for the Eurobalise Transmission System in Sect. 6.3.6, VBTS is composed of:

- trusted (safe) parts:
  - Virtual Balise Reader Functions,
  - TALS/Trackside Verification functions,
  - Railway Reference Stations of the Rail GNSS Augmentation Network (when the railway application requires it),

- Non trusted parts:
  - Global Navigation Satellite System, total system, i.e. the combined ground and airborne subsystems, in its role as a source of positioning errors (failures and feared events originating from the system),
  - Airgap as the set of interfaces among SVs and on-board train GNSS Antenna, among SVs and Railway Reference Station GNSS Antenna (when the railway application requires the use of the Rail GNSS Augmentation Network). Therefore, the airgap refers to the GNSS signal in space as a source of positioning errors (feared events originating from the propagation environment),
  - On-board GNSS antenna.

A feared event is any event described in Part I of the book that can lead to a balise location error greater than the on-board estimated balise location accuracy and, then, to hazardous consequences.

All the safety-related functions of the virtual balise transmission system must be compliant with the applicable CENELEC Standards such as EN 50126 [7], EN 50128 [10] and EN 50129 [5].

The Minimum Operational Performance Standards (MOPSs) associated with the new Virtual Balise Transmission System is expected to be defined in a Railway MOPS. This Railway MOPS will be one of the deliverables of the Shift2Rail Technological Demonstrator TD2.4 - Fail-Safe Train Positioning, see Sect. 6.5. In the context of this proposed architecture, Virtual Balises can be used only for replacing fixed balises and are expected to be mainly used for resetting the train confidence interval by the position/linking function implemented in the ERTMS/ETCS kernel. Moreover, VBs shall not be used to provide information to the on-board ERTMS/ETCS kernel that, if missed could result in a hazardous consequence; in most cases, the RBC can provide such information (e.g. TSR) to the on-board kernel by means of messages with acknowledgement. For the cases where the RBC cannot provide critical information to the on-board kernel, e.g. "Stop if in Staff Responsible" and "Danger for SH Information", physical balises must be used. However, the use of a smart algorithm for detecting virtual balises, see Sect. 7.4.3, may remove such a constraint.

This expected use does not imply operation limitation in many ERTMS operational scenarios.

As linking information and virtual balise locations in the on-board track description come from the same signalling track layout (i.e. this is not the case for physical balises because the location of a physical detected balise only depends on the actual location of the physical balise independently from the foreseen location in the signalling track layout), the signalling design process and the signalling data preparation process should be enhanced to also mitigate this lack of independence.

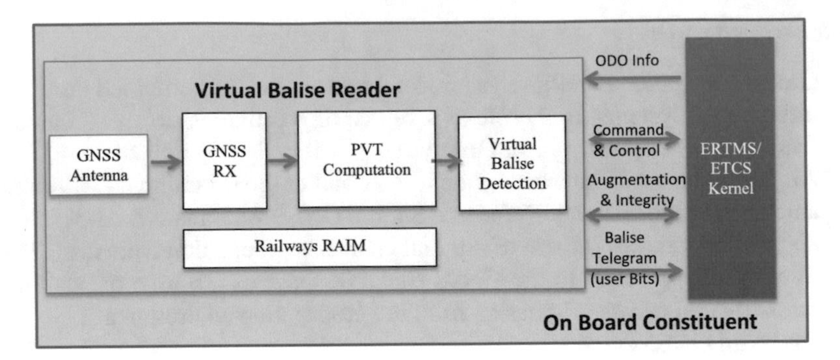

Fig. 7.3 High-level VBR architecture

7.4 The Virtual Balise Reader Functional Architecture

The Virtual Balise Reader (VBR) functional block comprises the core of the additional on-board functions needed to implement the virtual balise concept in the context of ERTMS/ETCS. It is composed of the GNSS Antenna, the GNSS Receiver Function, the PVT Computation and Virtual Balise Detector Function as well as the railway RAIM. The main inputs of the VBR are GNSS SIS signals, augmentation information, signalling information received from the trackside module TALS/Trackside Verification, and time and odometer reference from the ERTMS/ETCS kernel. The VBR computes both (a) unconstrained train position and related HPL and (b) constrained train position with related ATPL. It uses the constrained train position to carry out the detection of virtual balises. The VBR (a) computes continuously position information based on GNSS SIS and the augumentation information received and (b) compares it with a list of absolute reference positions stored in the on-board track description to detect the virtual balise. Once a virtual balise has been detected, the VBR delivers the balise information associated with the detected virtual balise, the time or odometer stamp for enabling the computation of the virtual balise center and the dynamically computed estimation of the virtual balise location accuracy. The VBR executes start-up tests to guarantee its normal behaviour at the railway mission start-up. Moreover, it also executes run-time tests to detect any failure of the virtual balise detection function and reports this failure event to the ERTMS/ETCS kernel.

A high-level description of the constituents of the VBR functional block as detailed in (Fig. 7.3) is provided in the following.

7.4.1 GNSS Antenna / GNSS RX

The "GNSS Antenna" is the radiating element that receives the GNSS SIS, while the "GNSS RX" includes the RF front-end with the ADC, processes digital signals and forwards different measurements to the "PVT Computation" functional block to implement railway specific PVT algorithm. Such measurements include but are not limited to pseudorange and carrier phase measurements, pseudorange residuals, navigation data and carrier-to-noise ratio. The "GNSS RX" functional block does not receive any railway signalling information (e.g. odometry, track description) nor any augmentation information.

7.4.2 PVT Computation

This functional block implements GNSS PVT algorithms specific to the railway environment; its main functions can be summarized as follows:

- Compute, when necessary (e.g. in SoM), the unconstrained position information with the corresponding horizontal protection level,
- Compute the constrained position information (i.e. the position already constrained on the track) with the corresponding ATPL by making use of the track description information; this track description information is dynamically updated by the trackside TALS module via specific new packets sent through Euroradio messages,
- Provide integrity to the computed position by using augmentation information and the output of the additional monitoring techniques implemented by the Railways RAIM. The combined use of the augmentation information and these monitoring techniques enables this functional block to compute a protection level taking into account the various sources of error that induce position uncertainty, for example, ionosphere and troposphere propagation, multipath or electromagnetic interference threats.
- Implement fault detection and exclusion algorithms using both GNSS information and railway signalling information: the purpose of these further defensive techniques is the detection of local-feared GNSS events that might lead to unbounded position errors in accordance with the assigned THR; for example, the use of SIL 4 odometry information based on the multi-sensor technology already available on the on-board has been demonstrated as a valid mitigation technique to any residual hazard associated with GNSS misleading information.

7.4.3 Virtual Balise Detector (VBD)

The VBD functional block carries out the following functions:

(a) Compare the GNSS-based train constrained position with pre-known virtual balise positions stored in the on-board track description to declare virtual balise detection. A smart algorithm for declaring the virtual balise detection also based on the SIL 4 odometry and its confidence interval can guarantee a hazard rate associated with the missed detection of a virtual balise much lower than the corresponding value for the BTM, $THR_{BTM-H1} = 1.0 \times 10^{-7}$ / h as described in Sect. 6.3.6 and enables the definition of a new less demanding mission profile.

(b) Receive odometry data from ERTMS/ETCS kernel to properly stamp the detected virtual balise,

(c) Provide the following information to the ETCS on-board kernel when a virtual balise is detected:

- Time / odometer stamp of the detected virtual balise centre; this information takes into account the position of the GNSS Antenna mounted on the train roof and projected on the track with respect to the front-end of the respective engine and with respect to the train orientation,
- The detection error associated with the virtual balise detection accuracy, see Sect. 7.5,
- Balise information (user bits) for the detected virtual balise, stored into an on-board track description,

(d) Guarantee the delivery of virtual balises in the correct sequence,

(e) Guarantee crosstalk protection in accordance with the assigned THR,

(f) Ensure immunity to environmental noise by means of the implementation of robust detection algorithms,

(g) Detect the corruption of user bits,

(h) Dynamically execute run-time tests to detect failures in the virtual balise detection function and notify this anomaly to the ERTMS/ETCS kernel. In particular, the use of both physical and virtual balise groups in the signalling trackside design might mean that some parameters of the standard reference mission profile, described in Sect. 6.3.5, may no longer be applicable. As the mission profile plays a critical role in the safety analysis, the completeness of the run-time tests and its impact on the mission profile require further analysis.

7.4.4 Railways RAIM

Due to the peculiarities of the railway environment with respect to RF channel impairments, see Sect. 6.6, the Railways RAIM functional block includes the set of cascaded integrity checks to be executed by the on-board VBR to cope with GNSS system and local feared events that may have an impact on the GNSS position to be used for detecting the virtual balise.

The first set of these integrity checks can be executed in the GNSS receiver at the level of the measured satellite signal individually. The objective of these checks is to isolate and exclude faulted or suspected code measurements and phase measurements before the next integrity checks at the level of PVT computation conduct the final check in both the range and the position domain. As the GNSS receiver is not a safe railway component, diagnostic information must also be provided to the PVT

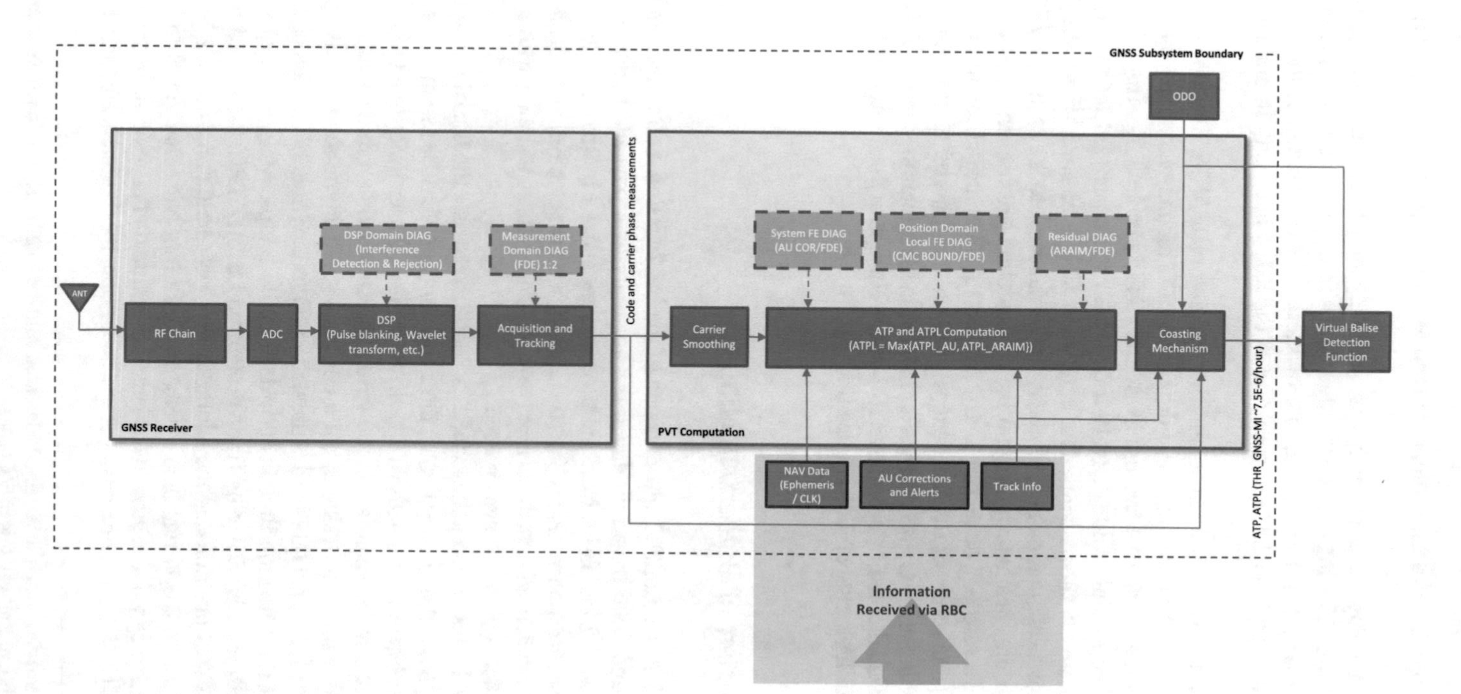

Fig. 7.4 Some details of the virtual balise reader

computation block (i.e. a safe component) so as it can implement safe periodic run-time checks.

For the remaining set of integrity checks, many studies are still in progress and different possible CRAIM or ARAIM algorithms have been proposed, see Part I of the book and [16, 17]. In addition, one of these integrity checks is the odometry cross check that exploits the high precision of the SIL 4 odometry in limited spatial intervals.

Finally, as far as the implementation of the complete VBR functional block is concerned, it must be compliant with the CENELEC recommendations about SIL 4 platform. Therefore, at least two **independent** GNSS channels, i.e. from the GNSS Antenna to the virtual balise delivery along with the correlated information, are required. Figure 7.4 shows some details of each GNSS channel where the GNSS receiver is not a safe component, whereas the remaining blocks such as PVT Computation, coasting mechanism, virtual balise detection are implemented on a safe railway architecture (e.g. on one replica of the 2oo2 architecture). In addition, Fig. 7.4 also outlines that some set of cascaded integrity checks (part of the Railway RAIM) are implemented in the GNSS receiver and others in the remaining blocks. This architecture has been defined by Ansaldo STS and the European Space Agency (ESA) in the context of their cooperation framework.

7.5 Accuracy of Balise Detection

Section 6.3.4 has described the concept of balise **location reference**, and both Sects. 6.3.4 and 6.3.6 have introduced the accuracy of a physical balise detection as the accuracy of the balise location reference **measured** by the on-board. Moreover, these sections have also reported the safe performance requirement for vital purposes associated with such accuracy for physical balises as being within ± 1 m, when a physical balise has been passed.

The virtual balise concept described in Sect. 7.2 enables the use of balise groups as done for physical balises and thus, the same concept of balise **location reference** is also applicable to virtual balises. For virtual balises, the accuracy of a virtual balise detection refers to the accuracy of the virtual balise location reference estimated by the on-board VBR. This value is not a constant value anymore, and it is dynamically estimated; it is the on-board estimated maximum virtual balise detection error computed at the detection time in accordance with the pre-assigned integrity; it is based on both the GNSS Along-Track Protection Level (ATPL) associated with the last valid[4] GNSS position and the odometry error accumulated from this last valid GNSS position to the virtual balise detection location. Note that the virtual balise detection accuracy value represents a statistical bound on the position error such that the probability of the position error being greater than this bound is less than or equal

[4]Valid GNSS Position means the position computed with the satellite signals that have successfully passed all on-board Railway Integrity Checks.

to the pre assigned integrity risk (see Sect. 7.4). The ERTMS/ETCS specifications allocate uncertainties associated with the balise detection accuracy as follows:

- over-bounding of inaccuracy in estimating relative position of the balise location reference, in the on-board odometric coordinate system, is in charge of the balise reader;
- over-bounding of inaccuracy in estimating absolute position of the balise reference location in the railway line (e.g., position with respect to signals or lineside indicators) is in charge of the trackside subsystem.

This association is evident when physical balises are used, as accuracy in installation of the balises is part of the collection, interpretation, accuracy and allocation of data relating to the railway network and the engineering of it into ETCS trackside data (i.e. dimensioning of the Q_LOCACC values), whereas the on-board is solely responsible for the accuracy in the balise detection.

In order to allow safe and available operational scenarios, to take into account the dynamic value of the virtual balise detection accuracy, and to minimize impact on current ERTMS/ETCS implementation when virtual balises are used, the following approach is proposed.

Signalling trackside engineers still consider the accuracy of the balise location reference (independently from the type of balise, physical or virtual) with respect to a target location and/or consecutive linked balise groups. Even though the uncertainty of the virtual balise position due to track constraints may not be relevant (e.g. a virtual balise is not physically mounted on a sleeper), there are measurement uncertainties related to the linking distances between virtual balise groups that must be taken into account independently from the precise surveying techniques used. In addition, let us remember that any distance information is evaluated on-board as nominal information (without taking into account any tolerances). Therefore, in order to enable on-board platform to take such uncertainty into account, the same variable Q_LOCACC adopted for the physical balise is proposed to be used for virtual balises. Furthermore, in order to reduce modification to the ERTMS/ETCS kernel for coping with (a) the dynamic value of the virtual balise location reference accuracy and (b) the concept of expectation window for virtual balises, a further important role has been proposed for the variable Q_LOCACC. For virtual balises, the Q_LOCACC constant value can be also dimensioned for including an a priori estimate of the virtual balise detection accuracy. As the balise detection accuracy plays a role in the interoperability among an ERTMS trackside CCS subsystem and ERTMS on-board CCS subsystems, this a priori estimate may be the minimum virtual balise detection inaccuracy that can be guaranteed for the interoperability. Based on the preliminary results obtained from the Italian Sardinia Trial Site [18] and the Australian Pre-Commissioning Field Tests, the value of first hypothesis for such a priori estimate can be set as:

A priori virtual balise detection accuracy $= 4\,\text{m} +$ average virtual balise detection accuracy $+7\times$ the related standard deviation $=$ about $20\,\text{m}$.

Where $4\,\text{m}$ is the fixed value included in the required space accuracy for the measurement of any travelled distance, see Sect. 6.3.4.

The preliminary ERTMS functional impact analysis based on such value (20 m) seems confirming no critical operation limitations. However, further detailed functional analysis will be carried out.

This proposed approach implies to split balise detection accuracy value in the following main components:

- accuracy related to the system of acquisition (i.e. accuracy of the balise reader, both for physical and virtual balises);
- accuracy related to the displacement of balises on the track (i.e., the Q_LOCACC value, that is an estimate of the location accuracy for the physical balises, and can be considered as an a-priori estimate of GNSS position error upper bound for virtual balises to guarantee the interoperability).

For physical balises, the balise reader estimates only the first component (that will be less than 1 m, as prescribed by ERTMS/ETCS specifications). For virtual balises, the balise reader estimates a value that embeds the two components (accuracy related to the system itself plus the GNSS position accuracy).

Therefore, for virtual balises, Q_LOCACC continues to be configured as a constant value and dimensioned as part of the trackside signalling design: its dimensioning must include both any uncertainties related to the trackside measurements and the a-priori estimate of the virtual balise detection accuracy. Based on the above considerations, when a virtual balise is detected, the virtual balise detection function of the VBR, described in Sect. 7.4, computes the detection error value associated with the virtual balise detection accuracy as follows:

Tot_Err = dynamically estimated virtual balise detection accuracy + the cumulative VBR uncertainties owing to the internal estimation process (excluding the a-priori estimate already included in Q_LOCACC)
if (Tot_Err > Q_LOCACC), then

detection error value = MAX (Tot_Err – Q_LOCACC, 1 m)

else

detection error value = 1 m

where 1 m is the nominal value that has been considered for the detection of a physical balise location, see Sect. 6.3.4. Please, let us remember that, when a virtual balise is detected, the VBD function provides such detection error value along with the following information to the ETCS on-board kernel:

- Time / odometer stamp of the detected virtual balise centre; this information takes into account the position of the GNSS Antenna mounted on the train and projected on the track,
- The **detection error** associated with the virtual balise detection accuracy,
- Balise information (user bits) for the detected virtual balise, stored into an on-board track description.

Given that the **detection error** value of the virtual balise is included in the computation of the train safe front-ends and of the train confidence interval, large detection error values do not have any impact on the safety (provided that the dynamically estimated virtual balise detection accuracy is always greater than the actual detection error in accordance with the pre assigned integrity). A large value has an impact on the operational intrusiveness only; the train will stop in advance of the supervised target location. Therefore, temporarily large values can be tolerated during the train missions.

7.6 Train Confidence Interval Based on Physical and Virtual Balises

Based on the virtual balise concept described in Sect. 7.2 and the ERTMS basic signalling principles outlined in Sect. 6.3.4, the **accuracy of ERTMS/ETCS train positioning** using both physical balises and virtual balises can logically remain unchanged and still be influenced by:

- the detection error associated with the accuracy of balise detection, i.e. 1 m for physical balises or the value dynamically estimated for virtual balises as described in Sect. 7.5,
- the accuracy of the balise location on the track or on the track description stored on-board (i.e. Q_LOCACC dimensioned as recommended in Sect. 6.3.4 and in Subset 040 [3] for physical balises or in Sect. 7.5 for virtual balises), and
- the accuracy of odometry (i.e. the accuracy of the distances measured on-board) that, for every measured distance s, shall be better or equal to (4 m + the accuracy of balise detection + 5 % s).

As described in Sect. 6.3.4, all location related information required for safe operation must be used by the on-board equipment taking into account the confidence interval of the train position. To this end, and with the use of also virtual balises, the ERTMS/ETCS kernel can continue to compute the train front-end position as follows (see Figs. 7.5 and 7.6):

(1) the estimated front-end position,
(2) the max(imum) safe front-end position, differing from the estimated position by the under-reading amount in the distance measured from the LRBG by odometry (composed of physical balises or virtual balises) plus both
    (a) the location accuracy of the LRBG (i.e. Q_LOCACC dimensioned as recommended in Sect. 6.3.4 and in Subset 040 [3] for physical balises or in Sect. 7.5 for virtual balises) and
    (b) the detection error associated with the virtual balise detection accuracy as described in Sect. 7.5. Please, note that, in relation to the orientation of the train, this position is in advance of the estimated position,
(3) the min(imum) safe front-end position, differing from the estimated position by the over-reading amount in the distance measured from the LRBG by odometry (composed of physical balises or virtual balises) plus both

(a) the location accuracy of the LRBG (dimensioned as for the maximum safe fronte end) and
(b) the detection error associated with the virtual balise detection accuracy as described in Sect. 7.5. Please, note that, in relation to the orientation of the train, this position is in rear of the estimated position.

As a consequence, the accuracy of the supervision of a generic target objective (e.g. EOA) can be computed by taking into account:

- the knowledge of the train position (with respect to the LRBG, composed of physical or virtual balises), depending on the accuracy of train position (computed as described above) ensured by the ERTMS on-board equipment and on the spacing of balise groups (design of trackside signalling),
- the knowledge of the distance to the target objective, depending on resolution and accuracy of distances given in the MA messages and on accuracy (i.e. Q_LOCACC dimensioned as described in Sect. 7.5) of a balise installation for a physical balise or of a virtual balise location in the track description stored on-board,
- the knowledge of line gradients, depending on resolution and accuracy of information in profile messages,
- the accuracy of the train speed measured by the ERTMS on-board equipment,
- the accuracy of braking model and parameters (i.e. accuracy of the modelling of the train).

The knowledge of this supervision accuracy enables trackside signalling designers to compute a-priori the maximum absolute error that would be expected at the target objective locations and thus guaranteeing the interoperability with trains equipped by different ERTMS suppliers.

The need to take into account the inaccuracies in the position of the LRBG (composed of physical or virtual balises) for safe operation is evident considering that:

- all location and profile data transmitted from the RBC (trackside interoperable component) refers to the location reference of the LRBG,
- the on-board measured distance by using the odometry refers to the location reference of the LRBG,
- the train confidence interval still refers to the inaccuracies of the LRBG location reference measured or estimated on-board.

Moreover, as the linking check is based on an expectation window that includes the location accuracies of both the LRBG and the expected balise group, wrong dimensioning of Q_LOCACC may cause on-board equipment loosing a balise group and eventually ordering to activate the brakes.

For example, let us suppose that the on-board VBD correctly detects and processes a balise group composed of virtual balises, sends the position report to the RBC, and updates the on-board LRBG (i.e. the detected balise group becomes the on-board LRBG). Based on the received position report and the status of the line, let us suppose that the RBC provides a Movement Authority (MA) up to a target objective (e.g. a signal) that the train shall not overpass; this MA includes the distance that can be travelled from the LRBG and lets suppose that this is 1 km; this means that the foreseen nominal distance between the LRBG and the target objective is 1 km. If the real train position was only 950 m from the target objective when the LRBG reference location was detected, the train shall not overpass the target objective if and only if Q_LOCACC + Detection Error is larger than 50 m, where detection error is the dynamically estimated value when the LRBG was detected.

Figures 7.5 and 7.6 illustrate the relationships between dynamically estimated virtual balise detection accuracy, Q_LOCACC (dimensioned as described in Sect. 7.5) and the Over/under-reading amount in determining the train confidence interval.

7.7 Preliminary Apportionment of the Tolerable Hazard Rate Related to ERTMS/ETCS Core Hazard Based on the Virtual Balise Concept

7.7.1 Definition of GNSS Integrity Risk in the Railway Context

It is worth dedicating particular attention to the concept of GNSS integrity in the railway context and make distinctions with respect to the integrity concept defined for civil aviation application domain. RTCA DO-245A [19] defines that

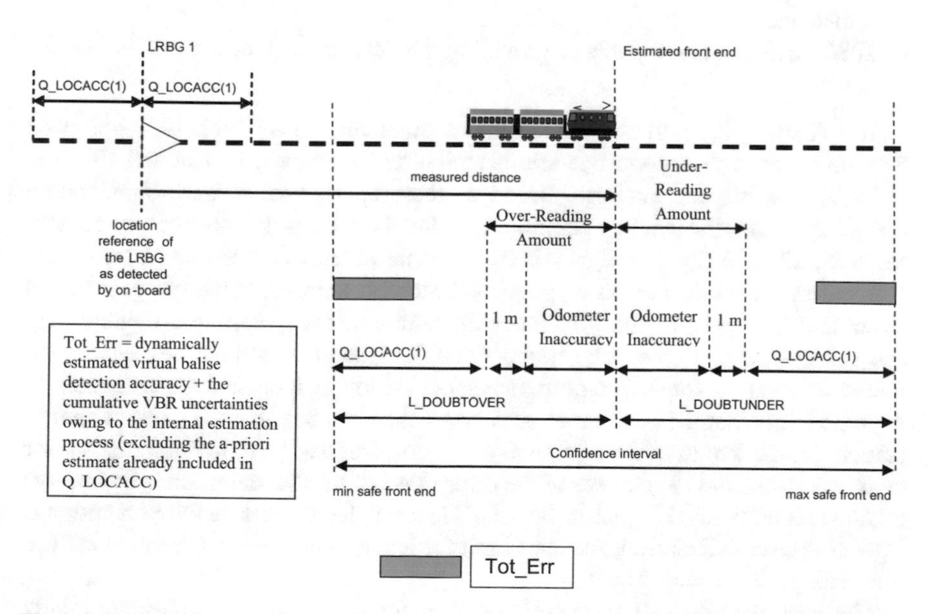

Fig. 7.5 Train confidence interval and train front-end position when Tot_Err ≤ Q_LOCACC

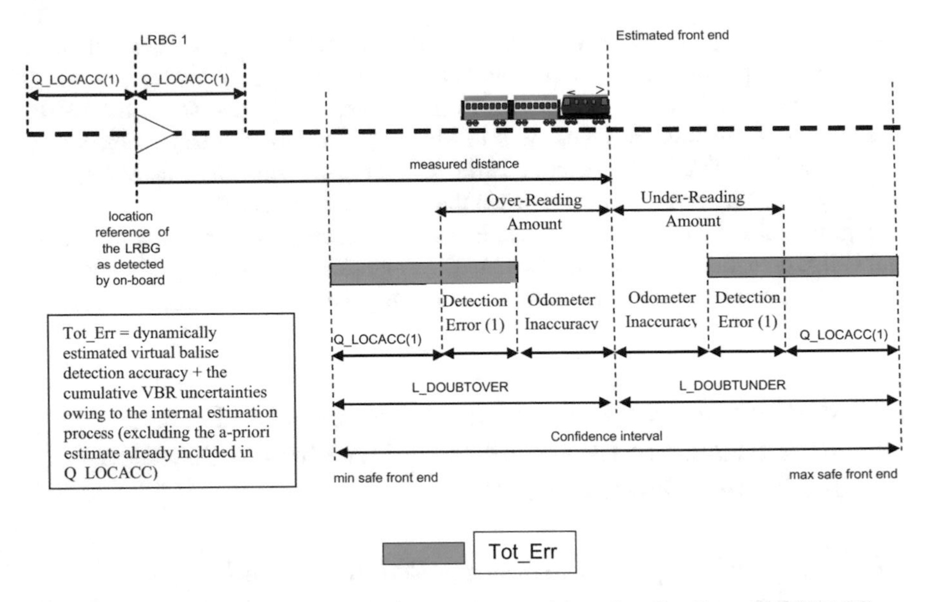

Fig. 7.6 Train confidence interval and train front-end position when Tot_Err > Q_LOCACC

- GPS/LAAS is intended to be used to provide radionavigation vertical and lateral guidance for aviation instrument flight rules (IFR) precision approach and landings from about 20 nautical miles (NM) runway threshold through touchdown and rollout, and
- GPS/LAAS must be capable of providing this service to all aircraft in the service volume.

RTCA DO-245A [19] also defines the requirements for a GPS/LAAS Approach Service intended to support operations including Category I, II, IIIa and IIIb precision approaches and landings. Based on these operational scenarios/application operational rules, the aviation community defined the integrity concept for aviation, see Sect. 1.2, as follows: Integrity is the measure of trust that can be placed in the correctness of the information supplied by the total system, the combined ground and airborne subsystems. Integrity includes the ability of the system to provide timely and valid warnings to the user (alerts) when the system should not be used for the intended operation. Integrity requirements for positioning include three elements: (1) the probability that the position error is larger than can be tolerated without annunciation, (2) the length of time (time to alert) the error can be out-of-tolerance prior to annunciation and (3) the size of the error (alert limit) that determines the out-of-tolerance condition. At Signal-in-Space (SIS) level, the out-of-tolerance condition is a position error that exceeds the alert limit for longer than the SIS time to alert. The true error position cannot be known.

However, the application operational rules for safe train positioning are totally different with respect to those of civil aviation and, therefore, the concept of integrity for railway operational environments should be defined for and agreed in the railway community. To this end, let us remember the key concept of Train Position Confidence Interval, defined in Sect. 6.3.4 for standard ERTMS systems: It is the distance interval within which the ERTMS/ETCS on-board assumes the actual train position is, with a defined probability. It comprises the odometer over-reading and under-reading amounts, plus twice the location accuracy of the reference balise group. The odometer over-reading and under-reading amounts also include the balise detection error. To cope with the concept of virtual balise, this definition has then been revised in Sect. 7.6 as follows: It is the distance interval within which the ERTMS/ETCS on-board assumes the actual train position is, with a defined probability. It comprises the odometer over-reading and under-reading amounts, plus twice the location accuracy of the reference balise group (composed of physical or virtual balises). The odometer over-reading and under-reading amounts also include the error contribution coming from the balise dynamic detection error estimated on-board at the virtual balise detection time. Please, note that ERTMS does not impose an upper bound on this train position confidence interval.[5]

The safe requirement associated with ERTMS Core Hazard, as defined in Sect. 6.3.6, requires that the actual train position must always be inside the train position confidence interval with a residual hazard rate equal to 10^{-9}/ h. Large train position confidence intervals have only an impact on some performance requirements such as availability, schedule adherence, intrusiveness, see Sect. 6.3.6. Please, note that railway signalling systems are designed to temporarily tolerate train position confidence intervals greater than those normally expected and these situations do not affect the safety.

Based on the above considerations and on the VBTS architecture outlined in Figs. 7.2, 7.3 and 7.4, the following definition of railway integrity is proposed: **Railway Integrity is the measure of trust that can be placed in the correctness of the GNSS Train Position information to be used to detect a virtual balise in the Virtual Balise Transmission System. The GNSS Train Position information is based on the GNSS information supplied by the total GNSS system, the combined ground and airborne subsystems. Railway Integrity includes the ability of the Virtual Balise Transmission System, based on GNSS Augmentation information, to provide timely and valid warnings to the ERTMS/ETCS kernel when the GNSS Train Position information should not be used for the intended operation (e.g. VBR is not able to detect any virtual balise due to its internal fault).**

Integrity requirements for train positioning include two elements: (1) the probability that the actual train position error for detecting a virtual balise is smaller than the maximum dynamically estimated GNSS train position error and (2) the size of the maximum dynamically estimated GNSS train position error (protection level) computed when the virtual balise detection occurs. At GNSS Signal in Space (SIS)

[5]There is only an upper value coming from the maximum range of the ERTMS variables used for the data format representation of such interval. However, this upper bound value is much higher than the admissible value for the typical railway operations.

level, the loss of railway integrity condition may lead to a GNSS train position error for detecting a virtual balise that exceeds the protection level computed at the virtual balise detection time (PE > (PL + VBR uncertainties owing to the internal estimation process)), i.e. this event is named GNSS misleading information. In addition, due to the nature and geometry of the railway environment, the protection level is usually computed based on a projection of the GNSS train position on the track where the train is supposed to be. For constrained GNSS train position, the loss of the GNSS railway integrity occurs when the GNSS Along-Track Position Error (ATPE) is greater than Along-Track Protection Level (ATPL). Note that the VBR does not declare unavailability if the protection level becomes greater than the nominal one. In addition, the true error position cannot be known.

The safe related parts of the Virtual Balise Transmission System must guarantee the railway integrity with respect to feared events originating from both the GNSS system (e.g. ephemeris errors, satellite failures, pseudorange errors, satellite clock errors and errors induced by the ionosphere) and the local environment (e.g. multipath, NLOS conditions, radio frequency interference). Furthermore, VBTS must also be developed and verified in accordance with safe and quality requirements recommended by the CENELEC Standards. In particular, these parts must be capable of continuing to meet their safety requirements in the event of random hardware faults, as far as reasonably practicable, systematic hardware or software faults, and GNSS related faults. Examples of GNSS-related faults are as follows:

- fault(s) in the GNSS Augmentation Network or in the Rail GNSS Augmentation Network or in the TALS/Trackside Verification module or in the on-board GNSS receiver or in the on-board GNSS Antenna or in the on-board PVT Computation module or in their combinations that leads (lead) to compute erroneous GNSS Train Position (the faulted case) for detecting a virtual balise, or
- fault(s) in one of the above-listed subsystems/modules or their combination that leads (lead) to compute unbounded or wrongly unbounded GNSS train position, not caused by a fault (fault-free case), for detecting a virtual balise.

Different safe architectures can be used to implement and demonstrate such a capability [5] with respect to safety requirements, see Figs. 7.7 and 7.8:

- Composite fail-safety—With this technique, each safety-related function is performed by at least two items. Each of these items shall be independent from all others, to avoid common-cause failures. Non-restrictive activities are allowed to progress only if the necessary number of items agree. A hazardous fault in one item shall be detected and negated in sufficient time to avoid a co-incident fault in a second item.
- Reactive fail-safety—This technique allows a safety-related function to be performed by a single item, provided its safe operation is assured by rapid detection and negation of any hazardous fault. Although only one item performs the actual safety-related function, the checking/testing/detection function shall be regarded as a second item, which shall be independent to avoid common-cause failures.

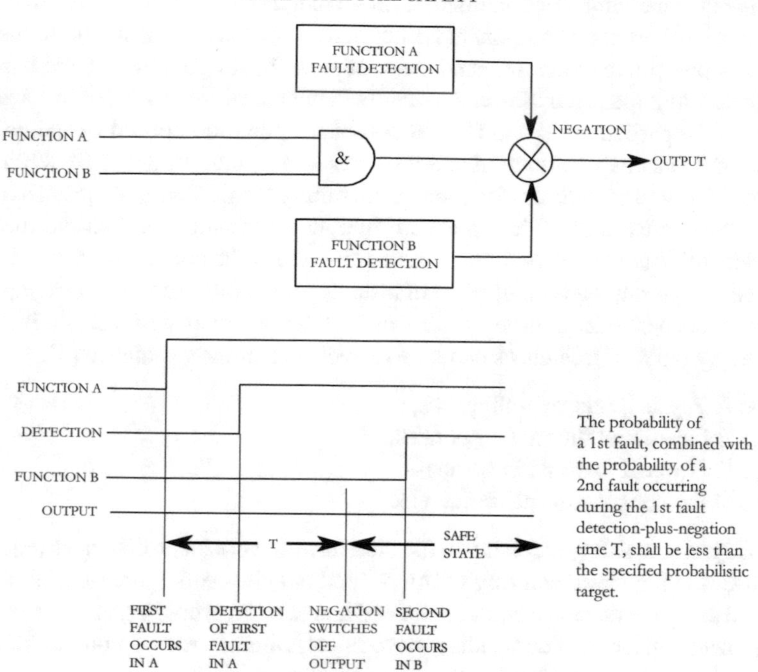

Fig. 7.7 Composite fail-safety

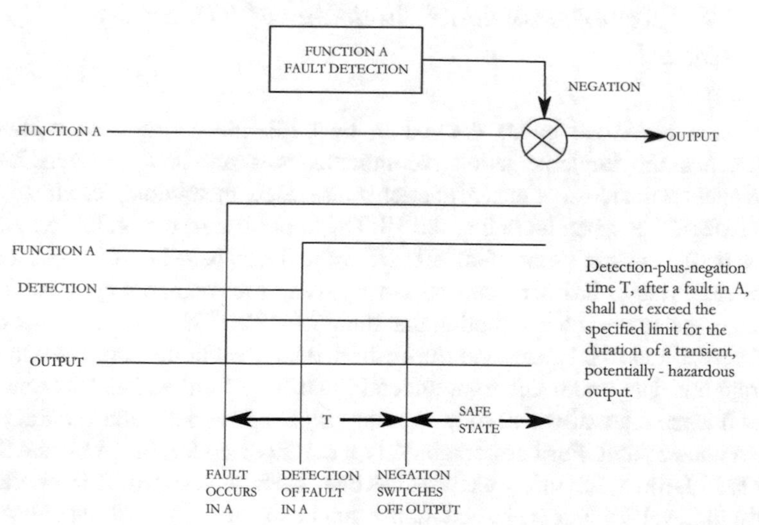

Fig. 7.8 Reactive fail-safety

Whichever technique or combination of techniques is used, assurance that (a) no single random hardware component failure mode is hazardous must be demonstrated by using appropriate structured analysis methods, (b) residual systematic hardware or software faults and GNSS-related faults cannot lead the VBTS into hazardous situations. In particular, as VBTS is a complex system composed of many items, which simultaneous malfunction could be hazardous, independence between items is a mandatory precondition for safety concerning single faults. Appropriate rules or guidelines must be fulfilled to ensure this independence, and that the measures taken/identified are effective for the whole life cycle of the complete system. In addition, the design must be arranged to minimize potentially hazardous consequences of loss of independence caused by, for example, a systematic design fault, if it could exist; many types of influences can lead to such loss of independence [5]:

- Type A Physical internal influences,
- Type B Functional internal influences,
- Type C Physical external influences,
- Type D Functional external influences.

Finally, where safety is reliant on the clearance and creepage distances (e.g. in the design and in the manufacturing of the printed circuit boards), the minimum clearance and creepage distances must be defined and verified according to the application requirements (in terms of demanding mechanical, climatic, radio frequency and electrical environments associated with rolling stocks).

7.7.2 THR Apportionment Methodology of ETCS Core Hazard

The concept of virtual balise defined in Sect. 7.2, the architecture outlined in Sect. 7.3, and the implementation recommendations provided in Sects. 7.4, 7.5 and 7.6 enables the reuse of many functional and safety methodologies already used for ERTMS/ETCS when including the VBTS. In particular, the preliminary safety analysis on the enhancement of an ERTMS system also based on the virtual balise transmission system has been carried out by using the methodology described in Sect. 6.3.6, and under the assumption that the ERTMS/ETCS kernel does not differentiate between virtual balises and Eurobalises other than in the computation of the train front-end, maximum safe front-end, minimum safe front-end and the train position confidence interval where the dynamic virtual balise detection accuracy must be taken into account. Furthermore, based on the expected use of the virtual balises for ERTMS Level 2, the safety analysis has been performed for ERTMS operational scenario in Level 2 where a physical balise can be replaced by a virtual balise.

7.7.3 *Initial Apportionment of ERTMS/ETCS Core Hazard*

Starting from the safety analysis performed for the Eurobalise Transmission System, see in Sect. 6.3.6 and the definition of the Virtual Balise Transmission System provided in Sect. 7.3, the following system-level transmission hazards have been identified:

- TRANS-BALISE-1 Incorrect balise group message received by on-board kernel functions as consistent (corruption),
- TRANS-BALISE-2 Balise group not detected by on-board kernel functions (Deletion),
- TRANS-BALISE-3 Inserted balise group message received by on-board Kernel functions as consistent (insertion/crosstalk).

Furthermore, in analogy with the case of physical balises, the safety analysis has been done considering both the inherent protections in ETCS against these hazardous events and the operational considerations for the use of virtual balises. Furthermore, the use of the virtual balises (as it has been prospected) must not change the maximum allowed rate of occurrence of the ETCS Core Hazard, this is 1.0×10^{-9} / h for ETCS on-board and 1.0×10^{-9} / h for ETCS trackside.

The contribution of TRANS-BALISE-1 is considered negligible with respect to TRANS-BALISE-2 and 3 because balise information associated with virtual balises (a) is transferred from the RBC to the on-board via the safe and secure ERTMS/ETCS communication session by means of EURORADIO channel and (b) is stored on-board on a railway safe platform that guarantees the data storage integrity required for SIL 4 railway applications. Therefore, considering the Corruption Hazard (TRANS-BALISE-1) negligible, the initial apportionment of 50% between the Deletion Hazard (TRANS-BALISE-2) and the Insertion Hazard (TRANS-BALISE-3) is considered. This gives an initial THR of, see Fig. 7.9:

TRANS-BALISE-2 = 3.3×10^{-10} dangerous failures per hour; and
TRANS-BALISE-3 = 3.3×10^{-10} dangerous failures per hour.

Assuming that (a) the RBC data preparation process for building and storing the balise information into the RBC database is suitable for SIL 4 railway applications, and (b) the uploading procedure for transferring the track description and the balise information from the RBC to the on-board is fault-free, the TRANS-BALISE-2 hazard originates only from the on-board. However, as the ERTMS/ETCS kernel can detect a missed balise (physical or virtual) only when linking information is available on-board, the safety analysis about the TRANS-BALISE-2 hazardous event must include both cases, with and without linking information.

TRANS-BALISE-2 with Linking

- When the linking information is available on-board, the on-board can detect if an expected balise group, composed of virtual balises, has been missed when the on-board passes the end of the expectation window. In the case of a missed detection, the linking reaction, if any, is commanded by the on-board.

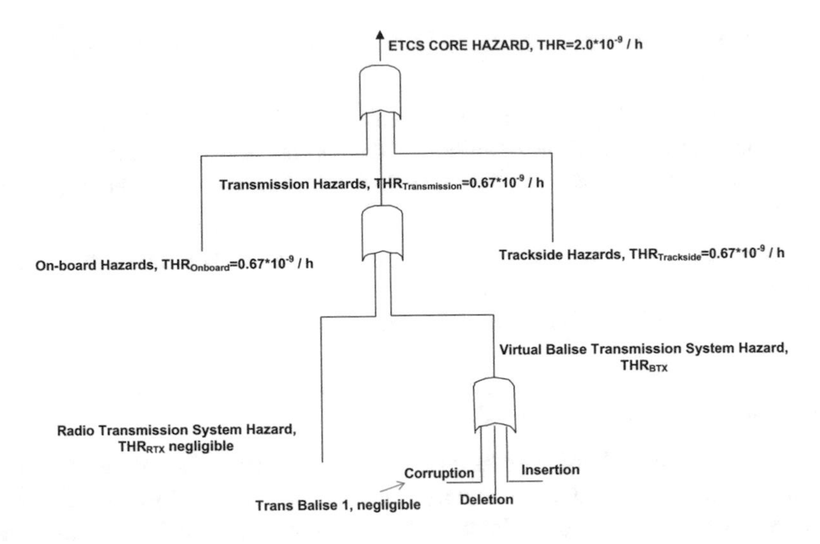

Fig. 7.9 ETCS core hazard apportionment with virtual balise among on-board, transmission and trackside

- When the virtual balise is detected, the balise information is always provided to the ERTMS/ETCS kernel. For virtual balise stored in a safe platform, the balise detection without balise information is a rare event because its probability of occurrence is the same as that of the VBR failure.
- A smart algorithm for declaring the virtual balise detection, also based on the SIL 4 odometry and its confidence interval, always guarantees the detection of a virtual balise provided that the PVT Computation Module can provide Constrained PVT (i.e. the GNSS PVT constrained on the track where the train is supposed to be).
- The balise group management policy of the ERTMS/ETCS kernel is remained unchanged; thus, independently of the linked reaction set by trackside, if two consecutive linked balise groups announced by linking information are not detected and the end of the expectation window of the second balise group has been passed, the ERTMS/ETCS on-board commands the service brake and the driver is informed. At standstill, the location -based information (i.e. location data, this is the set of the data that refer only to specific locations) stored on-board shall be shortened to the current position determined by the train odometry.

7.7.3.1 TRANS-BALISE-2 Without Linking

There is no inherent ETCS protection without linking; therefore, in order to mitigate potentially hazardous consequences resulting from missed balise detection, the following rules are proposed:

- All virtual balise groups shall be marked as “linked”.

- Virtual balises shall not be used to provide information to the ERTMS/ETCS kernel, that if missed, could result in a hazardous consequence.

However, the above-mentioned smart algorithm for declaring the virtual balise detection, also based on the SIL 4 odometry and its confidence interval, always guarantees the detection of a virtual balise provided that the PVT Computation Module can provide Constrained PVT (i.e. the GNSS PVT constrained on the track where the train is supposed to be). Therefore, based on the above considerations, the THR apportionment to TRANS-Balise-2 can be considered negligible.

7.7.3.2 TRANS-BALISE-3

Based on the above considerations, THR apportionment to TRANS-BALISE-3 becomes THRTRANS-BALISE-3 $= 6.7 \times 10^{-10}$ / h.

These preliminary results that have been reached are promising and will be the starting point for future *R&D* activities such the S2R TD2.4 Fail-Safe Train Positioning (including satellite technology) where:

- The functional requirements of the enhanced ERTMS also based on the GNSS position technology will be defined and agreed among the railway stakeholders;
- The detailed functional and non functional requirements of the virtual balise transmission system will be provided;
- The functional system hazard analysis will be carried out; and
- The railway VBTS minimum operational performance standards will be defined.

References

1. European GNSS Agency (GSA). E-GNSS in Rail Signalling Roadmap (2016). https://www.gsa.europa.eu/sites/default/files/futureflowchartd.pdf
2. European Space Agency (ESA). Space4Rail projects (2017). http://space4rail.esa.int/projects
3. UNISIG. SUBSET-040 v.3.4.0, Dimensioning and Engineering rules
4. UNISIG. SUBSET-036 v.3.1.0, FFFIS for Eurobalise
5. CENELEC. EN 50129 Railway Applications Communication, signalling and processing systems Safety related electronic systems for signalling
6. ERTMS Users Group. ERTMS/ETCS RAMS Requirements Specification Chapter 2 RAM, Reference EEIG: 96S126 V6, September 1998
7. CENELEC. EN 50126 Railway Applications. The specification and demonstration of reliability, availability, maintainability and safety (RAMS)
8. UNISIG. SUBSET-023 v.3.3.0, Glossary of Terms and Abbreviations
9. UNISIG. SUBSET-026 v3.6.0 ERTMS/ETCS System Requirements Specification
10. CENELEC. EN 50128 Railway Applications - Communication, signalling and processing systems - Software for railway control and protection systems
11. ERSAT-EAV project. ERTMS on SATELLITE Enabling Application and Validation (2017). http://www.ersat-eav.eu/
12. RHINOS project. Railway High Integrity Navigation Overlay System (2017). https://www.gsa.europa.eu/railway-high-integrity-navigation-overlay-system

13. Gianluigi Fontana and Salvatore Sabina. 3INSAT Demup - Demo Utilization Plan. Technical Report "S00A.0100402.P17.00EN", Ansaldo STS, August 2015
14. Barbara Brunetti, Salvatore Sabina, Giacomo Donati, Valerio di Claudio, and Pasquale Giugliano. 3INSAT CCN#2 Demup - Demo Utilization Plan. Technical Report "S00A.0100402.P17.01EN", Ansaldo STS, April 2016
15. CENELEC. EN 50159 Railway applications - Communication, signalling and processing systems - Safety-related communication in transmission systems
16. Hewitson S, Wang J (2006) GNSS receiver autonomous integrity monitoring (RAIM) performance analysis. GPS Solut 10(3):155–170
17. RHINOS project. D11.2 - Initial description of the Integrity Architecture. 2017
18. 3INSAT project. Train Integrated Safety Satellite System (2016). https://business.esa.int/projects/3insat
19. RTCA DO-245A. Minimum aviation system performance standards for the local area augmentation system (LAAS)

Appendix A
ERTMS/ETCS Railway Signalling

Salvatore Sabina, Fabio Poli and Nazelie Kassabian

A.1 Interoperable Constituents

The basic interoperability constituents in the Control-Command and Signalling Subsystems are, respectively, defined in Table A.1 for the Control-Command and Signalling On-board Subsystem [1] and Table A.2 for the Control-Command and Signalling Trackside Subsystem [1].

The functions of basic interoperability constituents may be combined to form a group. This group is then defined by those functions and by its remaining external interfaces. If a group is formed in this way, it shall be considered as an interoperability constituent. Table A.3 lists the groups of interoperability constituents of the Control-Command and Signalling On-board Subsystem [1]. Table A.4 lists the groups of interoperability constituents of the Control-Command and Signalling Trackside Subsystem [1].

S. Sabina (✉)
Ansaldo STS S.p.A, Via Paolo Mantovani 3-5, 16151 Genova, Italy
e-mail: salvatore.sabina@ansaldo-sts.com

F. Poli
Ansaldo STS S.p.A, Via Ferrante Imparato 184, 80147 Napoli, Italy
e-mail: fabio.poli@ansaldo-sts.com

N. Kassabian
Ansaldo STS S.p.A, Via Volvera 50, 10045 Piossasco Torino, Italy
e-mail: nazelie.kassabian@ansaldo-sts.com

© Springer International Publishing AG, part of Springer Nature 2018
L. Lo Presti and S. Sabina (eds.), *GNSS for Rail Transportation*, PoliTO Springer Series, https://doi.org/10.1007/978-3-319-79084-8

Table A.1 Basic interoperability constituents in the Control-Command and Signalling On-board Subsystem

1	2	3	4
N	Interoperability constituent IC	Characteristics	Specific requirements to be assessed by reference to Chap. 4 [1]
1	ETCS on-board	Reliability, Availability, Maintainability, Safety (RAMS)	4.2.1
			4.5.1
		On-board ETCS functionality (excluding odometry)	4.2.2
		ETCS and GSM-R air gap interfaces	4.2.5
		– RBC (level 2 and level 3)	4.2.5.1
		– Radio in-fill unit (optional level 1)	4.2.5.1
		– Eurobalise air gap	4.2.5.2
		– Euroloop air gap (optional level 1)	4.2.5.3
		Interfaces	
		– STM (implementation of interface K optional)	4.2.6.1
		– GSM-R ETCS Data Only Radio	4.2.6.2
		– Odometry	4.2.6.3
		– Key management system	4.2.8
		– ETCS ID Management	4.2.9
		– ETCS Driver–Machine Interface	4.2.12
		– Train interface	4.2.2
		– On-board recording device	4.2.14
		Construction of equipment	4.2.16
2	Odometry equipment	Reliability, Availability, Maintainability, Safety (RAMS)	4.2.1
			4.5.1
		On-board ETCS functionality (only Odometry) Interfaces	4.2.2
		– On-board ETCS	4.2.6.3
		Construction of equipment	4.2.16
3	Interface of External STM	Interfaces	
		– On-board ETCS	4.2.6.1

(continued)

Table A.1 (continued)

1	2	3	4
N	Interoperability constituent IC	Characteristics	Specific requirements to be assessed by reference to Chap. 4 [1]
4	GSM-R voice cab radio Note: SIM card, antenna, connecting cables and filters are not part of this interoperability constituent	Reliability, Availability, Maintainability, Safety (RAMS) Note: no requirement for safety	4.2.1
			4.5.1
		Basic communication functions	4.2.4.1
		Voice and operational communication applications	4.2.4.2
		Interfaces	
		– GSM-R air gap	4.2.5.1
		– GSM-R Driver–Machine Interface	4.2.13
		Construction of equipment	4.2.16
5	GSM-R ETCS Data only Radio Note: SIM card, antenna, connecting cables and filters are not part of this interoperability constituent	Reliability, Availability, Maintainability, Safety (RAMS) Note: no requirement for safety	4.2.1
			4.5.1
		Basic communication functions	4.2.4.1
		ETCS data communication applications	4.2.4.3
		Interfaces	
		– On-board ETCS – GSM-R air gap	4.2.6.2 4.2.5.1
		Construction of equipment	4.2.16
6	GSM-R SIM card Note: it is the responsibility of the GSM-R network operator to deliver to railway undertakings the SIM cards to be inserted in GSM-R terminal equipment	Basic communication functions	4.2.4.1
		Construction of equipment	4.2.16

Table A.2 Basic interoperability constituents in the Control-Command and Signalling Trackside Subsystem

1	2	3	4
N	Interoperability constituent IC	Characteristics	Specific requirements to be assessed by reference to Chap. 4 [1]
1	RBC	Reliability, Availability, Maintainability, Safety (RAMS)	4.2.1
			4.5.1
		Trackside ETCS functionality (excluding communication via Eurobalises, radio in-fill and Euroloop)	4.2.3
		ETCS and GSM-R air gap interfaces: only radio communication with train	4.2.5.1
		Interfaces	
		– Neighbouring RBC	4.2.7.1, 4.2.7.2
		– data radio communication	4.2.7.3
		– Key management system	4.2.8
		– ETCS-ID Management	4.2.9
		Construction of equipment	4.2.16
2	Radio in-fill unit	Reliability, Availability, Maintainability, Safety (RAMS)	4.2.1
			4.5.1
		Trackside ETCS functionality (excluding communication via Eurobalises, Euroloop and level 2 and level 3 functionality)	4.2.3
		ETCS and GSM-R air gap interfaces: only radio communication with train	4.2.5.1
		Interfaces	
		– data radio communication	4.2.7.3
		– Key management system	4.2.8
		– ETCS-ID Management	4.2.9
		– Interlocking and LEU	4.2.3
		Construction of equipment	4.2.16
3	Eurobalise	Reliability, Availability, Maintainability, Safety (RAMS)	4.2.1
			4.5.1
		ETCS and GSM-R air gap interfaces: only Eurobalise communication with train	4.2.5.2
		Interfaces	
		– LEU – Eurobalise	4.2.7.4
		Construction of equipment	4.2.16

(continued)

Table A.2 (continued)

1	2	3	4
N	Interoperability constituent IC	Characteristics	Specific requirements to be assessed by reference to Chap. 4 [1]
4	Euroloop	Reliability, Availability, Maintainability, Safety (RAMS)	4.2.1
			4.5.1
		ETCS and GSM-R air gap interfaces: only Euroloop communication with train	4.2.5.3
		Interfaces	
		– LEU – Euroloop	4.2.7.5
		Construction of equipment	4.2.16
5	LEU Eurobalise	Reliability, Availability, Maintainability, Safety (RAMS)	4.2.1
			4.5.1
		Trackside ETCS functionality (excluding communication via radio in-fill, Euroloop and level 2 and level 3 functionality)	4.2.3
		Interfaces	
		– LEU – Eurobalise	4.2.7.4
		Construction of equipment	4.2.16
6	LEU Euroloop	Reliability, Availability, Maintainability, Safety (RAMS)	4.2.1
			4.5.1
		Trackside ETCS functionality (excluding communication via radio in-fill, Eurobalise and level 2 and level 3 functionality)	4.2.3
		Interfaces	
		– LEU – Euroloop	4.2.7.5
		Construction of equipment	4.2.16

Table A.3 Groups of interoperability constituents in the Control-Command and Signalling On-board subsystem

1	2	3	4
N	Interoperability constituent IC	Characteristics	Specific requirements to be assessed by reference to Chap. 4 [1]
1	ETCS on-board Odometry Equipment	Reliability, Availability, Maintainability, Safety (RAMS)	4.2.1
			4.5.1
		On-board ETCS functionality	4.2.2
		ETCS and GSM-R air gap interfaces	4.2.5
		– RBC (level 2 and level 3)	4.2.5.1
		– Radio in-fill unit (optional level 1)	4.2.5.1
		– Eurobalise air gap	4.2.5.2
		– Euroloop air gap (optional level 1)	4.2.5.3
		Interfaces	
		– STM (implementation of interface K optional)	4.2.6.1
		– GSM-R ETCS Data Only Radio	4.2.6.2
		– Key management system	4.2.8
		– ETCS ID Management	4.2.9
		– ETCS Driver-Machine Interface	4.2.12
		– Train interface	4.2.2
		– On-board recording device	4.2.14
		Construction of equipment	4.2.16

Table A.4 Groups of interoperability constituents in the Control-Command and Signalling Track-side Subsystem

1	2	3	4
N	Interoperability constituent IC	Characteristics	Specific requirements to be assessed by reference to Chap. 4 [1]
1	Eurobalise LEU Eurobalise	Reliability, Availability, Maintainability, Safety (RAMS)	4.2.1
			4.5.1
		Trackside ETCS functionality (excluding communication via Euroloop and level 2 and level 3 functionality)	4.2.3
		ETCS and GSM-R air gap interfaces: only Eurobalise communication with train	4.2.5.2
		Construction of equipment	4.2.16
2	Euroloop LEU Euroloop	Reliability, Availability, Maintainability, Safety (RAMS)	4.2.1
			4.5.1
		Trackside ETCS functionality (excluding communication via Eurobalise and level 2 and level 3 functionality)	4.2.3
		ETCS and GSM-R air gap interfaces: only Euroloop communication with train	4.2.5.3
		Construction of equipment	4.2.16

A.2 Cross-Talk Protected Zone

To define a cross-talk protected zone, let us first provide the definition of the following coordinate system used for the orientation of the Balise and the Antenna Unit of the Eurobalise Transmission System with respect to the rail direction; see Fig. A.1:

- A reference axis in parallel with the rails (the X-axis),
- A reference axis at right angles across the rails, and which is level with the top of rails (the Y-axis),
- A reference axis directed upwards, at right angles to the rail plane (the Z-axis).

The Balise has reference marks on each of the six sides that indicate the positions of the three axes, related to the electrical centre of the Balise. The Antenna Unit has reference marks on each of the six sides that indicate the positions of the X-, Y-, and Z-axes, respectively.

Based on such reference axis, Table A.5 provides the definition of the cross-talk protected zone. The complete description of intrinsic cross-talk protection for Eurobalise Transmission System is provided in [2].

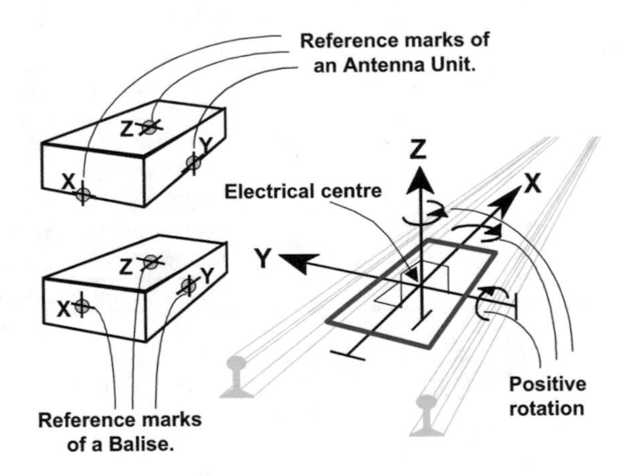

Fig. A.1 Eurobalise Transmission System reference axes

Table A.5 Definition of cross-talk protected zone

Type of cross-talk	Involved equipment	Zone where cross-talk shall not occur
Lateral (direction Y)	One Balise and one antenna unit	1.4 m or more between the Balise and the antenna unit (related to the Z reference marks)
Lateral (direction Y)	One or two Balises and two antenna units	3.0 m or more between the cross-talk Balise and the interfered antenna unit (related to the Z reference marks)
Vertical (direction Z)	One Balise and one antenna unit	4.8 m or more talk Balise related to the X and Y reference marks
Longitudinal (direction X)	Two Balises and one antenna unit. 2.6 or more between two consecutive standard size Balises, and 2.3 or more between two reduced size Balises (related to the Y reference marks). 2.6 m applies if combinations of Balise sizes are applicable	Any location of the antenna unit along the same track as the Balises
Longitudinal (direction X)	One Balise and two antenna units. 4.0 or more between two antenna units	Any location of the antenna units along the same track as the Balises

References

1. EU. Commission Regulation (EU) 2016/919 of 27 May 2016 on the technical specification for interoperability relating to the control-command and signalling subsystems of the rail system in the European Union, 27 May 2016
2. UNISIG. SUBSET-036 v.3.1.0 (2016) FFFIS for Eurobalise

Glossary

Automatic Train Protection A safety system that enforces either compliance with or observation of speed restrictions and signal aspects by trains.

Balise Group One or more Balises which are treated as having the same reference location on the track. The telegrams transmitted by all the Balises of a group form a track-to-train message.

Balise Group Coordinate System The means to ensure common location referencing between on-board and trackside, for all location-based information exchanged through the ERTMS/ETCS transmission media.

Balise Group Location reference Location of Balise number 1 in a Balise group. It is the origin of the Balise group coordinate system.

Expectation window The interval between the outer limits to accept a Balise group.

Fail-safe A design philosophy which results in any expected failure maintaining or placing the equipment in a safe state.

Last Relevant Balise Group The LRBG is used as a common location reference between the ERTMS/ETCS on-board and trackside equipments in levels 2 and 3.

Level The different ERTMS/ETCS application levels are a way to express the possible operating relationships between track and train. Level definitions are related to the trackside equipment used, to the way the trackside information reaches the on-board units and to which functions are processed in the trackside and in the on-board equipment, respectively.

Linking A functionality to protect against missing data from BALISE GROUPS by announcing them in advance through LINKING INFORMATION and by checking whether they have been read within a certain EXPECTATION WINDOW.

Mission Any train movement started under the supervision of an ERTMS/ETCS on-board equipment in one the following modes: FS, LS, SR, OS, NL, UN or SN. The ETCS mission is ended when any of the following modes is entered: SB, SH.

© Springer International Publishing AG, part of Springer Nature 2018

L. Lo Presti and S. Sabina (eds.), *GNSS for Rail Transportation*, PoliTO Springer Series, https://doi.org/10.1007/978-3-319-79084-8

Mode An operating state of the ERTMS/ETCS on-board equipment with a specified split of operational responsibilities between the ERTMS/ETCS system and the driver.

Movement Authority Permission for a train to run to a specific location within the constraints of the infrastructure.

Moving block A block whose length is defined by the position of the train occupying the section of track ahead. The minimum block length would be from the rearmost part of the occupying train to a point on the track where, if the train braked from its current speed, the front of the occupying train would be when the train came to a stand.

Radio Block Centre A centralised safety unit that receives train position information via radio and sends movement authorities via radio to trains.

Spot Transmission Transmission between trackside and on-board that takes place at discrete locations.

Track description Information complementing the Movement Authority and providing as a minimum the static speed profile and gradient profile. Optionally, it can contain axle load profile, track conditions, route suitability data, areas where shunting is permitted, etc.

Train data Defined set of data which gives information about the train. Data that characterises a train and which is required by ERTMS/ETCS in order to supervise a train movement.

Train detection The proof of the presence or absence of trains on a defined section of line.

Train integrity The level of belief in the train being complete and not having left coaches or wagons behind.

Train movement When vehicles are moved with train data available, as a rule from station to station, and as a rule under the authority of proceed aspects from main signals, or similar procedures.

Train position information The train position information defines the position of the train front in relation to a Balise group, which is called Last Relevant Balise Group (LRBG).

Printed in the United States
By Bookmasters